NUREG-1852

Demonstrating the Feasibility and Reliability of Operator Manual Actions in Response to Fire

Final Report

Manuscript Completed: September 2007
Date Published: October 2007

Prepared by
A. Kolaczkowski/Science Applications Internal Corporation
J. Forester/Sandia National Laboratories
R. Gallucci, A. Klein/U.S. Nuclear Regulatory Commission
J. Bongarra, P. Qualls, P. Barbadoro/U.S. Nuclear Regulatory Commission

Sandia National Laboratories
P.O. Box 5800
Albuquerque, NM 87185

Science Applications International Corporation
405 Urban Street, Suite 400
Lakewood, CO 80220

E. Lois, NRC Project Manager

Prepared for:
Division of Risk Assessment and Special Projects
Office of Nuclear Regulatory Research
U.S. Nuclear Regulatory Commission
Washington, DC 20555-0001

ABSTRACT

This report provides criteria and associated technical bases for evaluating the feasibility and reliability of postfire operator manual actions implemented in nuclear power plants. The U.S. Nuclear Regulatory Commission (NRC) developed this report as a reference guide for agency staff who evaluate the acceptability of manual actions, submitted by licensees as exemption requests from the requirements of Paragraph III.G.2 of Appendix R, "Fire Protection Program for Nuclear Power Facilities Operating Prior to January 1, 1979," to Title10, Part 50, "Domestic Licensing of Production and Utilization Facilities," of the *Code of Federal Regulations* (10 CFR Part 50), as a means of achieving and maintaining hot shutdown conditions during and after fire events. The staff may use this information in the review of *future* postfire operator manual actions to determine if the feasibility and reliability of the operator manual action were adequately evaluated.

FOREWORD

This report provides criteria and associated technical bases for evaluating the feasibility and reliability of postfire operator manual actions implemented in nuclear power plants. The U.S. Nuclear Regulatory Commission (NRC) developed this report as a reference guide for agency staff who evaluate the acceptability of manual actions, submitted by licensees as exemption requests from the requirements of Paragraph III.G.2 of Appendix R, "Fire Protection Program for Nuclear Power Facilities Operating Prior to January 1, 1979," to Title10, Part 50, "Domestic Licensing of Production and Utilization Facilities," of the *Code of Federal Regulations* (10 CFR Part 50), as a means of achieving and maintaining hot shutdown conditions during and after fire events. The staff may use this information in the review of *future* postfire operator manual actions to determine if the feasibility and reliability of the operator manual action were adequately evaluated. The work was performed by the NRC's Office of Nuclear Regulatory Research and Office of Nuclear Reactor Regulation, with support from Sandia National Laboratories and its contractor.

This report was developed on the basis of NRC and contractor experience in evaluating plans at nuclear power plants for human performance during fire events (e.g., inspections of plants' fire protection programs) and the review of work related to modeling human behavior in response to fires and other accident conditions in nuclear power plants. Reviewed documents include, but are not limited to, fire analyses conducted as part of individual plant examinations of external events (IPEEEs), the IPEEE summary report (NUREG-1742, "Perspectives Gained from the Individual Plant Examination of External Events [IPEEE] Program," Volumes 1 and 2, issued April 2002), fire-related operational events, the fire requantification work conducted jointly by the NRC and the Electric Power Research Institute (EPRI) (NUREG/CR-6850 [EPRI TR-1011989], "EPRI/NRC-RES Fire PRA Methodology for Nuclear Power Facilities," issued September 2005), and the American National Standards Institute/American Nuclear Society Standard 58.8-1994, "American National Standard Time Response Design Criteria for Safety-Related Operator Actions."

The technical information provided in this report is aimed at providing assistance to agency staff when they need to evaluate that postfire operator manual actions are both feasible and reliable. In addition, the information may be useful to licensees choosing to use such an approach to demonstrate the feasibility and reliability of operator manual actions. Among the criteria provided is the importance of time-authenticated demonstrations of the manual actions (involving actual execution of the actions to the extent possible) and adequate time available to complete the actions before fire-induced consequences occur that would otherwise prevent achieving and maintaining hot shutdown.

This report focuses on *unique* aspects of the hazard involved (fire), as well as the potentially unique characteristics of subsequent manual actions during the operators' response. Hence, it does not address all the various facets of programs that could potentially impact human performance during a fire. For instance, this report does not specify in detail what constitutes "adequate procedures"; other guidance documents already address this issue. Nonetheless, this report addresses the unique aspects of fire and associated operator manual actions to guide the NRC staff in determining whether operator manual actions, proposed by operating plants for use in achieving and maintaining hot shutdown, are feasible and can reliably be performed in response to fire.

Farouk Eltawila, Director
Division of Risk Assessment & Special Projects
Office of Nuclear Regulatory Research
U.S. Nuclear Regulatory Commission

CONTENTS

Page

ABSTRACT. iii

FOREWORD. v

ABBREVIATIONS. xi

GLOSSARY. xiii

1. INTRODUCTION. 1-1

2. DISCUSSION. 2-1
    2.1 Background. 2-1
    2.2 Purpose of this Report. 2-3
    2.3 Scope of this Report. 2-4

3. BASES FOR THE FEASIBILITY AND RELIABILITY CRITERIA. 3-1
    3.1 Overview. 3-1
    3.2 Summary of Bases for Feasibility and Reliability Criteria. 3-3
        3.2.1 Analysis Showing Adequate Time Available to Perform the Actions
                (To Address Feasibility). 3-3
        3.2.2 Analysis Showing Adequate Time Available to Ensure Reliability
        3-3
        3.2.3 Environmental Factors. 3-6
        3.2.4 Equipment Functionality and Accessibility. 3-7
        3.2.5 Available Indications. 3-8
        3.2.6 Communications. 3-9
        3.2.7 Portable Equipment. 3-10
        3.2.8 Personnel Protection Equipment. 3-11
        3.2.9 Procedures and Training. 3-11
        3.2.10 Staffing. 3-13
        3.2.11 Demonstrations. 3-15

4. TECHNICAL INFORMATION FOR IMPLEMENTING THE FEASIBILITY
   AND RELIABILITY CRITERIA. 4-1
    4.1 Overview. 4-1
    4.2 Technical Information for Feasibility and Reliability Criteria. 4-1
        4.2.1 Information Regarding the Analysis Showing Adequate Time
                Available to Perform the Actions to Address Feasibility. 4-1
        4.2.2 Information Regarding the Analysis Showing Adequate Time
                Available to Ensure Reliability. 4-3
        4.2.3 Information Regarding Environmental Factors. 4-6
        4.2.4 Information Regarding Equipment Functionality and Accessibility. 4-7
        4.2.5 Information Regarding Available Indications. 4-8
        4.2.6 Information Regarding Communications. 4-9

    4.2.7 Information Regarding Portable Equipment. 4-10

    4.2.8 Information Regarding Personnel Protection Equipment. 4-10

    4.2.9 Information Regarding Procedures and Training. 4-11

    4.2.10 Information Regarding the Staffing Criterion. 4-13

    4.2.11 Information Regarding How to Perform a Demonstration. 4-14

5. REFERENCES.. 5-1

APPENDIX A
    GUIDELINES FOR USING TIMELINES TO DEMONSTRATE SUFFICIENT TIME TO
    PERFORM THE ACTIONS. A-1

APPENDIX B
    SUMMARY OF EXPERT OPINION ELICITATION TO DETERMINE TIME MARGINS
    FOR OPERATOR MANUAL ACTIONS IN RESPONSE TO FIRE
    (April 1–2 and May 4–5, 2004). B-1

    B.1 Introduction. B-1

    B.2 First Expert Elicitation Meeting. B-2

        B.2.1 Expert Panel and Qualifications. B-2

        B.2.2 Summary of Topics Discussed During the First Meeting. B-3

        B.2.3 Example Elicitation Cases Addressed at the First Meeting. B-9

        B.2.4 Conclusion from First Meeting. B-13

    B.3 Second Expert Elicitation Meeting. B-13

        B.3.1 Summary of Topics Discussed During the Second Meeting.. . . . . B-14

        B.3.2 Determination of Time Margin. B-18

    B.4 Identification of Time Margin and Conclusion.. . . . . B-22

    B.5 Characteristics of an Expert Elicitation Panel. B-23

    B.6 References.. . . . . B-23

Figures

                                                               Page

A-1   A timeline. A-1

A-2   Initial fire detection and multiple action (one action dependent on a
           separate diagnosis of an undesired equipment failure) with a
           single overall time margin and T_3. . . . . A-3

A-3   Initial fire detection and multiple actions (one action dependent on
           completion of a prior action) with a single overall time margin and T_3. . . . . A-4

A-4   Initial fire detection and multiple actions illustrating the application of multiple
           time margins and T_3s.. . . . . A-5

B-1   Conceptual illustration of a time margin.. . . . . B-3

Tables

                                                                          Page

B-1   Initial and Revised Additional Times Added to Combined T_1 and T_2. . . . . B-11

B-2   Initial Time Added for Diagnosing the Need and Successfully Closing Open PORVs. . . B-13

B-3   Total Time for Each Step of the Action for the Third Scenario,
        by Panel Member (Base Time Plus Time Added for Influence Factors). B-20

B-4   Time Added to Each Step of the Manual Action for the Fourth Scenario
        (Hybrid Case of a Preventive and a Reactive Action). B-22

ABBREVIATIONS

ANS   American Nuclear Society
ANSI   American National Standards Institute
ATHEANA   A Technique for Human Event Analysis

BWR   boiling-water reactor

CFR   *Code of Federal Regulations*
CCW   component cooling water

EOP   emergency operating procedure
EPRI   Electric Power Research Institute
ESWGR   East Switchgear Room

GL   generic letter

HRA   human reliability analysis

IN   information notice
IPEEE   individual plant examination of external events

LOCA   loss-of-coolant accident

MCR   main control room
MOV   motor-operated valve

NEI   Nuclear Energy Institute
NFPA   National Fire Protection Association
NRC   U.S. Nuclear Regulatory Commission
NRR   Office of Nuclear Reactor Regulation (NRC)

PEO   plant equipment operator
PORV   power-operated relief valve
PWR   pressurized-water reactor

RES   Office of Nuclear Regulatory Research (NRC)
RIS   regulatory issue summary

SCBA   self-contained breathing apparatus
SPAR-H   Simplified Plant Analysis Risk—Human Reliability Analysis
SSC   structure, system, and component

WSWGR   West Switchgear Room

GLOSSARY

Below are key terms or phrases whose definitions and associated context, for purposes of this document, are as shown.

action—An activity, typically observable, and usually involving the manipulation of equipment, that is carried out by an operator(s) to achieve a certain outcome. The required diagnosis of the need to perform the activity, the subsequent decision to perform the activity, obtaining any necessary equipment, procedures, or other aids or devices necessary to perform the activity, traveling to the location to perform the activity, implementing the activity, and checking that the activity has had its desired effect, are all implied and encompassed by the term "action."

available time (or time available)—The time period from a presentation of a cue for an action to the time of adverse consequences if the action is not taken.

diagnosis time—The time required for an operator(s) to examine and evaluate data to determine the need for, and to make the decision to implement, an action.

feasible action—An action that is analyzed and demonstrated as being able to be performed within an available time so as to avoid a defined undesirable outcome. As compared to a reliable action (see definition), an action is considered feasible if it is shown that it is possible to be performed within the available time (considering relevant uncertainties in estimating the time available); but it does not necessarily demonstrate that the action is reliable. For instance, performing an action successfully one time out of three attempts within the available time shows that the action is *feasible*, but not necessarily reliable.

implementation time—The time required by the operator(s) to successfully perform the manipulative aspects of an action (i.e., not the diagnosis aspects themselves, but typically as a result of the diagnosis aspects), including obtaining any necessary equipment, procedures, or other aids or devices; traveling to the necessary location; implementing the action; and checking that the action has had its desired effect.

*operator manual action*s (local actions, in response to a fire)—Those actions performed by operators to manipulate components and equipment from outside the main control room to achieve and maintain postfire hot shutdown, but not including "repairs." Operator manual actions comprise an integrated set of actions needed to help ensure that hot shutdown can be accomplished, given that a fire has occurred in a particular plant area.

preventive actions—Those actions that, upon entering a fire plan/procedure, the operator(s) takes (without needing further diagnosis) to mitigate the potential effects of possible spurious actuations or other fire-related failures, so as to ensure that hot shutdown can be achieved and maintained. For these actions, it is generally assumed that once the fire has been detected and located, per procedure, the control room crew will direct personnel to execute a number of actions, possibly even without the existence of other damage symptoms, to ensure the availability of equipment to achieve its function during the given fire scenario. In many cases, the only criterion for initiating these actions is the presence of the fire itself.

reactive actions—Those actions taken during a fire in response to an undesired change in plant condition. In reactive actions, the operator(s) detects the undesired change and, with the support of procedural guidance, diagnoses the correct actions to be taken. Thus, with reactive actions, the plant staff responds to indications of changing equipment conditions caused by the fire, and then takes the steps necessary to ensure that the equipment will function when needed (e.g., manually reopening a spuriously closed valve). The plant staff may not initiate the actions until the procedure indicates that, given the relevant indications, the actions must be performed.

reliable action—A feasible action that is analyzed and demonstrated as being dependably repeatable within an available time, so as to avoid a defined adverse consequence, while considering varying conditions that could affect the available time and/or the time to perform the action. As compared to an action that is only feasible (see definition), an action is considered to be reliable as well if it is shown that it can be dependably and repeatably performed within the available time, by different crews, under somewhat varying conditions that typify uncertainties in the available time and the time to perform the action, with a high success rate. All reliable actions need to be feasible, but not all feasible actions will be reliable.

1. INTRODUCTION

The primary objective of fire protection programs at U.S. nuclear plants are to minimize the effects of fires and explosions on structures, systems, and components (SSCs) important to safety. To meet this objective, fire protection programs for operating nuclear power plants are designed to provide reasonable assurance, through defense-in-depth, that (1) a fire will not prevent the performance of necessary safe shutdown functions, and (2) radioactive releases to the environment in the event of a fire will be minimized.

To provide those assurances, at least in part, many plants plan to or already rely on local operator manual actions[1] (i.e., actions outside the main control room (MCR)) to maintain hot shutdown capability. That is, operators either take preventive, local manual actions upon detecting a fire to protect critical safety equipment that might be failed or spuriously affected and rendered unavailable by the fire, or they locally and manually align critical safety equipment to perform its function when needed. Paragraph III.G.1 of Appendix R, "Fire Protection Program for Nuclear Power Facilities Operating Prior to January 1, 1979," to Title 10, Part 50, "Domestic Licensing of Production and Utilization Facilities," of the *Code of Federal Regulations* (10 CFR Part 50) [Ref. 1] states that one train of equipment needed to maintain hot shutdown conditions shall be free of fire damage. Paragraph III.G.2 of Appendix R specifies the following three methods, any of which are acceptable, to provide reasonable assurance that at least one means of achieving and maintaining hot shutdown conditions will remain available during and after any postulated fire in the plant[2], when redundant trains of equipment required for hot shutdown are in the same fire area outside of the primary containment:

(1) separation of redundant trains by a fire barrier having a 3-hour rating

(2) separation of redundant trains by a horizontal distance of more than 6.1 meters (20 feet) containing no intervening combustible or fire hazards, together with fire detectors and an automatic fire suppression system

(3) separation of redundant trains by a barrier having a 1-hour rating, coupled with fire detectors and an automatic fire suppression system.

If any one of the above cannot be met, then Paragraph III.G.3 requirements must be met. Operator manual actions can be used to satisfy Paragraph III.G.1 [Ref. 1] requirements since these areas do not contain redundant safe shutdown trains. Operator manual actions are allowed to satisfy requirements in Paragraph III.G.3 in the performance of alternate or

---

[1] "Operator manual actions" are defined in the Glossary of this report. For this report, they do not include any actions within the MCR or the action(s) associated with abandoning the MCR in the case of a fire. Further, while the April 2001 edition of Regulatory Guide 1.189, "Fire Protection for Operating Nuclear Power Plants," had details on what constitutes hot shutdown for pressurized-water reactors (PWRs) and boiling-water reactors (BWRs), including the required systems, Revision 1 of Regulatory Guide 1.189, "Fire Protection for Nuclear Power Plants," issued March 2007 [Ref. 6], excludes that discussion and just identifies the Technical Specifications of each plant providing the definitions of hot shutdown and cold shutdown. This document is applicable to only those actions to achieve and maintain hot shutdown.

[2] Similar guidance is incorporated into Section 9.5.1 of NUREG-0800, "Standard Review Plan for the Review of Safety Analysis Reports for Nuclear Power Plants," Revision 4, issued October 2003 [Ref. 2], for plants licensed after January 1, 1979. These "post-1979" licensees incorporate their fire protection program implementation requirements into their operating license as a license condition and those requirements are largely the same as those from Appendix R that are discussed throughout this report.

dedicated shutdown activities. The NRC proposed rulemaking in SECY 03-0100, "Rulemaking Plan on Post-Fire Operator Manual Actions," issued June 2003 [Ref. 3], states that, under certain circumstances, operator manual actions may be a reasonable alternative to the separation requirements of Paragraph III.G.2, and many operator actions for operation of a hot shutdown train during a fire would not involve any safety-significant concerns.

The NRC developed Regulatory Issue Summary (RIS) 2006-10, "Regulatory Expectations with Appendix R, Paragraph III.G.2, Operator Manual Actions," dated June 30, 2006 [Ref. 4], which discusses acceptable means for achieving compliance with 10 CFR 50.48, "Fire Protection" [Ref.5]. Although the title is specific to Appendix R [Ref.1], the RIS applies to plants that were licensed to operate both prior and subsequent to January 1, 1979. Therefore, this report provides criteria for demonstrating the feasibility and reliability of operator manual actions in response to fire that are applicable to all plants. The NRC staff recognizes that certain criteria must be met to ensure that adequate safety is maintained as a result of the use of operator manual actions as an alternative to separation/protection. In particular, the NRC staff notes that such actions must be both feasible and reliable, especially considering that these actions are relied upon in lieu of passive fire barriers, distance, separation, and/or fire detectors and automatic fire suppression systems, each with relatively high reliability.

This document provides technical bases in the form of criteria and related technical information for justifying that operator manual actions are feasible and can reliably be performed under a wide range of plant conditions that an operator might encounter during a fire. If a large number of manual actions need to be addressed, it is expected that for many cases, where extra time is clearly available and the actions are relatively simple, evaluating the criteria will be straightforward, requiring only simple justifications and analysis. Furthermore, for the complex cases, licensees alternatively may choose to comply with the requirements of Appendix R [Ref. 1] by performing appropriate design changes. For these cases, the licensees have the option of submitting an exemption or license amendment request using detailed analyses of operator manual action feasibility and reliability.

This report, as a reference guide, addresses the feasibility and reliability of operator manual actions, from a deterministic approach, when used to achieve and maintain hot shutdown under fire conditions. It is planned that this document will be used by the NRC staff to support the review of operator manual actions, submitted by licensees as exemption requests. However, an operator manual action which meets the information provided in this report does not necessarily comply with NRC fire protection regulations. Additional considerations to ensure that adequate defense-in-depth such as fire detection and automatic suppression is maintained are addressed in Regulatory Guide 1.189 [Ref. 6] and should be considered when applying for an exemption or license amendment.

Operator manual actions, allowed by existing regulations to satisfy Paragraphs III.G.1 and III.G.3 of Appendix R to 10 CFR Part 50 [Ref. 1], are often identical actions and need to have the same feasibility and reliability goals as the Paragraph III.G.2 operator manual actions which require prior staff approval to ensure an adequate level of plant safety. Many operator manual actions used for alternative or dedicated shutdown have received prior staff review and approval. This report will be used as information to review feasibility and reliability of *future*[3] postfire operator manual actions when staff review is required or requested.

Section 2 of this report explains the use of operator manual actions to ensure postfire hot shutdown, and discusses the purpose and scope of this report.

Section 3 summarizes each criterion, and discusses the basis for each.

Section 4 provides additional discussion of each criterion, as well as technical information for meeting the criteria.

---

[3] Throughout this document words are *italicized* for emphasis

2. DISCUSSION

2.1 Background

This section provides a brief summary of key historical events leading to the need to address the acceptability of certain postfire operator manual actions and, ultimately, the information provided in this report.

Title 10, Paragraph 50.48, of the *Code of Federal Regulations* [Ref.5], requires each operating nuclear power plant to have a fire protection plan that satisfies Criterion 3, "Fire Protection," of Appendix A, "General Design Criteria for Nuclear Power Plants," to 10 CFR Part 50 [Ref. 8]. Criterion 3 requires that SSCs important to safety must be designed and located to minimize, consistent with other safety requirements, the probability and effect of fires and explosions. The specific fire protection requirements for hot shutdown capability of a plant are further discussed in Section III.G of Appendix R to 10 CFR Part 50 [Ref. 1]. The more specific 10 CFR 50.48 [Ref. 5] and Appendix R requirements were added following a significant fire that occurred in 1975 at the Browns Ferry Nuclear Power Plant. That fire damaged control, instrumentation, and power cables for redundant trains of equipment necessary for hot shutdown.

In response to the fire, an NRC investigation revealed that the independence of redundant equipment at Browns Ferry was negated by a lack of adequate separation between cables for redundant trains of safety equipment. The investigators subsequently recommended that a suitable combination of electrical isolation, physical distance, fire barriers, and fixed automatic fire suppression systems should be used to maintain the independence of redundant safety equipment. In response to that recommendation, the NRC interacted with stakeholders for several years to identify and implement necessary plant fire protection improvements. In 1980, the NRC promulgated 10 CFR 50.48 [Ref. 5] to establish fire protection requirements and Appendix R to 10 CFR Part 50 [Ref. 1] for certain generic fire protection program issues, including Section III.G, which addresses fire protection of hot shutdown capability. The requirements for separation of cables and equipment associated with redundant hot shutdown trains within a fire area were promulgated in Paragraph III.G.2 of Appendix R for situations where fire area separation was not feasible (i.e., for plants already built or already designed).

Paragraph III.G.2 requires that cables and equipment of redundant trains of safety systems in the same fire area must be separated by one of the following provisions:

- a 3-hour fire barrier

- a horizontal distance of more than 6.1 meters (20 feet) with no intervening combustibles in conjunction with fire detectors and an automatic fire suppression system

- a 1-hour fire barrier combined with fire detectors and an automatic fire suppression system

Because the rule was to apply to facilities that were already built, the NRC realized that compliance with various parts of Appendix R might be difficult for certain fire areas such as the MCR and Cable Spreading Room at some facilities. Accordingly, the NRC included Paragraph III.G.3 to allow plants to credit dedicated or alternative safe shutdown equipment. There was also the provision to submit an exemption to seek NRC review and approval of alternative acceptable methods for protecting safe shutdown. During implementation of the requirements of Appendix R, the NRC reviewed and approved a large number of exemptions for 60 licensees, including numerous exemptions from Paragraphs III.G.2 and III.G.3.

In the early 1990s, generic problems arose with Thermo-Lag[4] fire barriers, which many licensees were using as either a 1- or 3-hour fire barrier to comply with Paragraph III.G.2 of Appendix R. As a result, the NRC ultimately required plants to upgrade existing Thermo-Lag electrical raceway fire barrier systems or provide another means of compliance with Appendix R. Several years later, however, fire protection inspectors began identifying instances where some plants had not upgraded or replaced the Thermo-Lag fire barrier material or provided the required separation distance between redundant safety trains used to satisfy the criteria of Paragraph III.G.2. Some plants compensated for this by relying on operator manual actions, which were not reviewed and approved by the NRC through the exemption process established by 10 CFR 50.12, "Specific Exemptions" [Ref. 9]. Nonetheless, the NRC recognized that such actions may be an acceptable way of achieving hot shutdown in the event of a fire under certain conditions.

In 2002, the NRC informed the nuclear power industry that the use of unapproved manual actions was not in compliance with Paragraph III.G.2. During a meeting on June 20, 2002, the Nuclear Energy Institute (NEI) representative stated that there was widespread use of operator manual actions throughout the industry based on the understanding of past practice and existing NRC guidance. The industry representative also stated that the use of unapproved manual actions had become prevalent even before the concerns arose with Thermo-Lag material. Subsequent to the public meeting, the NRC developed criteria for inspectors to use in assessing the safety-significance of violations resulting from use of unapproved operator manual actions. Those criteria were based on past practice and experience by NRC inspectors reviewing operator manual actions to comply with Paragraph III.G.3 on alternative reactor shutdown capability. Plant staff members were familiar with these criteria through their interactions with the NRC staff during the implementation of the NRC inspection process. These criteria were issued in the revision to Inspection Procedure 71111.05, "Fire Protection," in March 2003 [Ref. 10].

Building on the above inspection criteria, the NRC considered it prudent to codify criteria for licensees and the NRC staff to use in evaluating the acceptability of operator manual actions used in lieu of meeting the separation criteria in Paragraph III.G.2 of Appendix R where redundant trains of safety systems exist in the same fire area. These criteria were to ensure that the actions were both feasible and reliable. These criteria would maintain safety by ensuring that thorough evaluations of the operator manual actions were performed comparable to evaluations for an exemption request. A rule change to incorporate these criteria was begun but later abandoned on the basis that the desired effect of significantly reducing the number of exemption requests seeking approval of the use of these operator manual actions would not be realized.

Nevertheless, the identified criteria continue to be valid for evaluating the acceptability of these operator manual actions. Contemplating numerous exemption requests by licensees seeking

---

[4]   Thermo-Lag is a brand name for a particular type of material used to construct fire barriers typically for protecting electrical conduits and cable trays. In the early 1990s, issues arose regarding the testing and qualification process used for this material. It was determined that barriers made of this material would not provide protection for the required periods of time.

approval for such actions, the NRC staff considered it important to document these criteria and related information for use by the staff when evaluating the exemption requests. To meet that need, this report provides additional technical information, primarily for NRC staff, but also considered useful to the industry, for ensuring the feasibility and reliability of operator manual actions, from a deterministic approach, for post-fire hot shutdown.

2.2 <u>Purpose of this Report</u>

Most of the criteria provided herein are based on reviews of existing work related to modeling human behavior in responses to fires and other accident conditions in nuclear power plants. For example, most of the factors covered by the criteria were derived from reviews of selected fire analyses conducted as part of individual plant examinations of external events (IPEEEs), the IPEEE summary report (NUREG-1742, "Perspectives Gained From the Individual Plant Examination of External Events (IPEEE) Program," Volumes 1 and 2, issued April 2002 [Ref. 11]), previous reviews of fire-related operational events to identify important factors influencing human performance in fires [e.g., Refs. 12–14], lessons learned from the development of human reliability analysis (HRA) criteria for use in the ongoing fire requantification studies jointly conducted by the NRC and the Electric Power Research Institute (EPRI) [Ref. 15], general HRA methods such as Simplified Plant Analysis Risk—Human Reliability Analysis (SPAR-H) [Ref. 16] and A Technique for Human Event Analysis (ATHEANA) [Ref. 17], and information on operator response times and time response design criteria for safety-related operator actions [e.g., Refs. 18 and 19]. Examples of the general factors covered by the criteria (discussed in detail in Sections 3 and 4 of this report) include the availability of indications for the actions, environmental considerations, staffing and training, communications, availability of necessary equipment, and availability of procedures.

While the importance of such factors is generally obvious, determining exactly how to implement and evaluate the factors can be somewhat less straightforward and subject to interpretation. For example, what should be covered by procedures appropriate for operator manual actions, and what type of training is appropriate? One of the main purposes of this document is to provide additional technical information related to the factors as a means to address the acceptability of postfire manual actions using a deterministic approach.

This technical information is aimed at ensuring that operator manual actions are both feasible and reliable. Among the criteria provided herein is the need for time-authenticated demonstrations of the manual actions (involving actual execution of actions to the extent possible) and adequate time available to complete the actions before fire-induced consequences occur that would otherwise prevent achieving and maintaining hot shutdown. Showing, with a demonstration (as subsequently discussed in Sections 3 and 4 of this report), that actions meeting the criteria can be completed in the available time, documents the feasibility; however, additional issues must be considered to show that the actions can reliably be performed, by different crews, under the variety of conditions that could occur during a fire.

For example, fire factors which may not be possible to create for the demonstrations could cause further delay under actual fire conditions (i.e., the demonstration would likely fall short of actual fire situations). Hence, although a demonstration shows that a manual action can be performed within the necessary time, shortcomings of the demonstration may mean that under an actual fire situation, other possible delays not addressed by the demonstration may need to be accounted for. Furthermore, typical and expected variability between individuals and crews could lead to variations in operator performance (human-centered factors). Finally, variations in the characteristics of the fire and related plant conditions could alter the time available for the operator actions.

Hence, to ensure that actions can be performed reliably, and in concert with the safety margin philosophy inherent in the NRC's regulations as well as good engineering practice, the technical information provided herein also addresses the subject of performing analyses (or providing equivalent justification) useful to confirming that adequate time is available for the actions. The information strives to ensure that relevant factors are considered in determining or justifying time adequacy (which can be justified in different ways as subsequently addressed in Section 4 of this report) and that the process for determining the time available for the actions addresses the potential variations in fire characteristics and plant conditions.

As to the aforementioned analysis, and as delineated in greater detail in subsequent sections, determining whether there is enough time available to perform the operator manual action should account for potential circumstances, such as (1) the potential need to recover from or respond to unexpected difficulties associated with instruments or other equipment, or communication devices, (2) environmental and other effects that are not easily replicated in a demonstration, such as radiation, smoke, toxic gas effects, and increased noise levels, (3) limitations of the demonstration to account for all possible fire locations that may lead to the need for such operator manual actions, (4) inability to show or duplicate the operator manual actions during a demonstration because of safety considerations while at power, and (5) individual operator performance factors, such as physical size and strength, cognitive differences, and the effects of stress and time pressure. The time available should not be so restrictive relative to the time needed to perform the actions that personnel are not able to recover from any initial slips or errors in conducting the actions (i.e., there is some "recovery" time built in, should it be needed). Establishing that adequate time is available is more easily justified using demonstrations of the operator manual actions with clear illustration that appropriate calculations of the time available have been conducted. Sections 3 and 4 of this report provide further details regarding what should be considered in substantiating that adequate time is available to ensure the reliability of the operator manual actions.

As a final note, this report specifically addresses a deterministic approach for assessing the feasibility and reliability of operator manual actions. However, risk assessment and particularly human reliability techniques may be useful when identifying the range of fire scenarios and related contexts as well as the possible operator manual actions that might be used in response to possible fire scenarios. Hence, the use of such risk-related techniques as an aid in addressing the criteria presented herein is not discouraged, but the use of these techniques is neither required nor expected. Ultimately, using this information, the operator manual actions should meet the applicable deterministic criteria provided herein for feasibility and reliability.

2.3 Scope of this Report

This report provides technical information to assist the NRC staff in determining that operator manual actions are feasible and can be performed reliably in response to fire. The readers should refer to Regulatory Guide 1.189 [Ref. 6] for details on how the NRC staff plans to use this information in its reviews.

While this report strives to provide enough information to support this determination about the feasibility and reliability of manual actions, it does not attempt to cover in detail all possible aspects of how to meet the criteria that are provided herein. This report focuses on *unique* aspects of the hazard involved (fire) and the potentially unique characteristics of subsequent manual actions during the operators' response. Hence, for instance, it is not the intent of this report to specify in detail what constitutes "adequate procedures." Many other guidance documents and an evolving consensus address this issue. Additionally, each plant has a well-established program for identifying, writing, reviewing, issuing, and changing procedures. What is provided here is information on the unique aspects of fire and the associated operator manual actions.

Finally, for purposes of this report, the two types of operator manual actions covered by this technical information are (1) *preventive* actions and (2) *reactive* actions, as defined in the Glossary of this report.

3. BASES FOR THE FEASIBILITY AND RELIABILITY CRITERIA

3.1 Overview

This section presents the criteria for evaluating the feasibility and reliability of operator manual actions. Each criterion is briefly introduced, and the bases that support the need for each criterion are also provided. Technical information for implementing each criterion is covered in Section 4 of this report.

The following provides a definition (also in the Glossary) of "operator manual actions" as the term is used herein:

> Operator manual actions are those actions performed by operators to manipulate components and equipment from outside the MCR to achieve and maintain postfire hot shutdown, but do not include "repairs." Operator manual actions comprise an integrated set of actions needed to help ensure that hot shutdown can be accomplished, given that a fire has occurred in a particular plant area.

The NRC's feasibility and reliability criteria for operator manual actions are summarized below:

- An analysis should be prepared to evaluate the feasibility and reliability of operator manual actions. The *analysis* should determine that adequate time exists for the operator to perform the required manual actions to achieve and maintain hot shutdown from a single fire. The *adequate time* should reasonably account for all important variables, including (1) differences between the analyzed and actual conditions, and (2) human performance uncertainties that may be encountered.

- The analysis should show that the actions can be performed under the expected *environmental factors* that will be encountered.

- The analysis should show that (1) the *functionality of equipment and cables* needed to implement operator manual actions to achieve and maintain hot shutdown will not be adversely affected by the fire, and (2) the equipment will be *available* and readily *accessible* consistent with the analysis. In addition to the SSCs needed to directly perform the desired actions, other supporting equipment may also be required, including (to the extent required for successful performance of each operator manual action)—

  - *indications* necessary to show the need for the manual actions, enable their performance, and verify their successful accomplishment (if not directly observable)

  - necessary *communications*

  - necessary *portable equipment*

  - necessary *personnel protection equipment*

- There should be plant *procedures* covering each operator manual action required to achieve and maintain hot shutdown and *training* for each operator on the procedures.

- The *number of available personnel (staffing)*, exclusive of Fire Brigade members, needed to perform the actions should be consistent with the analysis.

- There should be periodic *demonstrations* of the manual actions, consisting of actual executions of the relevant actions to the extent sufficient to show continued proficiency in performing the actions.

The above criteria provide a means to determine that there is reasonable assurance that the actions are feasible and can be performed reliably to bring the plant to a hot shutdown condition, thereby protecting public health and safety. The above criteria are considered appropriate to achieve the overall requirement that the actions should be both feasible and reliable.

Any analysis of the feasibility and reliability of an operator manual action, or combination of such actions, begins with definition of the time available to complete the action(s). Typically, this dictates the degree of rigor needed for the rest of the analysis and the extent to which the various criteria need to be addressed. Since determination of the time to accomplish the action is inherent in developing the complete timeline, when accounting for fire effects, the analyst or reviewer will need to address the criteria that are applicable. However, not all of the criteria will usually require significant analysis or even be applicable, particularly for the simpler and more straightforward actions.

Suppose it was determined, based, for example, on an evaluation of the physical response of the plant to the transient that is potentially induced by the fire, that there is sufficient time available to complete an action (e.g., several hours). Assuming there are no "unique" aspects of the fire that could prolong its extinguishment unduly (e.g., an oil fire that is continuously fed by an oil leak, such that extinguishment is impossible until the leak can be stopped), and the proposed operator manual actions can be shown to be relatively straightforward and easily justified as unimpeded by the fire or its effects, including firefighting activities, one would expect that the various feasibility and reliability criteria can be shown to be met with very simple justifications or analysis. Some may not even be applicable, for example, there may be no need for special tools or "staged" equipment. Under such circumstances, one would expect a relatively simple analysis and review. Implicit would be an expectation of minimal variability and uncertainty along with a large "time margin." In such a case, for example, demonstration of the action may be easily justified as enveloped by some other action that is more rigorously demonstrated.

At the other extreme, suppose the time available is relatively short (tens of minutes, at most), the operator manual actions are not straightforward or are somewhat complex (or may involve multiple operators or the same operator performing multiple actions), or there are "unique" aspects to the fire making rapid extinguishment difficult (e.g., a reluctance to apply water to an electrical fire due to personnel safety considerations, such as the potential for high-voltage shocks[5]). Under such conditions, much more rigorous analysis and review are likely to be needed to account for all the criteria and how well each is met given the fire and its effects, including firefighting activities. It is certainly conceivable that, given the inherently greater variability and uncertainty present in such a situation, along with the inevitably short, if any,

---

[5] While this does not a priori include traditional reluctance to apply water to electrical fires, even when no personnel safety concern is involved, it should still be recognized that there may be such reluctance among some personnel, especially those not members of the Fire Brigade, such that time delays in extinguishment could be involved.

"time margin," that a deterministic evaluation using the criteria in this NUREG is not possible (due to the limits of the "all-or-none" nature of deterministic analysis).

The following subsections elaborate on the bases for each of the feasibility and reliability criteria. It should be noted that, in some cases, the various regulations and documents (e.g., NUREG-series reports) that are discussed below to provide a basis for the criteria are not necessarily tied directly to regulations that apply to specific plants, including the "pre-1979" plants (i.e., plants licensed to operate before January 1, 1979). The intent of the discussions, and the associated citations, is to illustrate that there is a defensible basis for why the various criteria are appropriate (i.e., the various factors and conditions they address represent sound practices that have already been identified as generally important to safety).

3.2 Summary of Bases for Feasibility and Reliability Criteria

3.2.1 Analysis Showing Adequate Time Available to Perform the Actions (To Address Feasibility)

This criterion addresses the need for an analysis to determine that there is adequate time available for the operator to perform the required manual actions to achieve and maintain hot shutdown after a single fire. The analysis should determine that the time available is long enough to allow the action to be diagnosed and executed. If a demonstration of the action (discussed in Section 3.2.11 below) shows that it can be diagnosed and accomplished in the time available, and uncertainties in estimating the time available have been considered (see Section 4.2.1 of this report), then the action can be regarded as feasible to achieve and maintain hot shutdown. To establish reliability, however, the uncertainties associated with estimating how long it takes to diagnose and execute operator manual actions (see Sections 3.2.2 and 4.2.2 of this report) should also be considered.

This criterion is based upon regulations requiring that a nuclear power plant must always be maintained in a safe condition, even following accidents, consistent with the additional restriction that a hot shutdown state be reached and maintained, in accordance with Section III.G of Appendix R to 10 CFR Part 50 [Ref. 1]. Implicit in these requirements is the analysis of the plant's thermal-hydraulic response, including the time needed to fulfill the listed safety functions.

This criterion is not a new NRC staff view in that previous NRC staff reviews and approvals of postfire operator manual actions included the consideration of whether there was adequate time for the operator manual actions, based on the progression of the fire and the thermal-hydraulic conditions of the plant. Additionally, this criterion is consistent with current inspection criteria for fire protection manual actions [Ref. 10] under the verification and validation criterion, ensuring that plant staff has adequately evaluated the capability of operators to perform the manual actions in the time available. These existing practices and the associated expectations support the need for a criterion that addresses the assurance that there is adequate time to perform the operator manual actions.

3.2.2 Analysis Showing Adequate Time Available to Ensure Reliability

This criterion addresses the reliability of the operator manual actions. For a feasible action to be performed reliably, it should be shown that there is adequate time available to account for uncertainties not only in estimates of the time available, but also in estimates of how long it takes to diagnose and execute the operator manual actions (e.g., as based, at least in part, on a plant demonstration of the action under nonfire conditions). It should be shown that there is extra time available to account for such uncertainties. This extra time is a surrogate for directly accounting for sources of uncertainty, such as the following, inherent in estimating the time available for the action and the time required:

(1)   variations in fire and related plant conditions that could affect the time estimates (e.g., fast energetic fire failing equipment quickly vs. slow developing fire with little or no equipment failures for some time, variable fire detector response times and sensitivities, variable air flows affecting the fire and its growth, specific fire initiation location relative to important targets, presence (or not) of temporary transient combustibles)

(2)   factors unable to be recreated in the demonstrations, or in some cases not anticipated for an actual fire situation, that could cause further delay in the time it could take to perform the operator manual actions under actual fire conditions (i.e., where the demonstration may likely fall short of actual fire situations), as in the following examples:

  –   The operators may need to recover from/respond to unexpected difficulties, such as problems with instruments or other equipment (e.g., locked doors, a stiff handwheel, or difficulty with communication devices). Such difficulties can and sometimes do happen and represent a possible uncertainty in how long it will take to perform an action. The extra time would make it unlikely that difficulties encountered in an actual fire situation will prevent the desired manual actions from being accomplished in a timely manner.

  –   Environmental and other effects might exist that are not included as part of the simulation in the demonstration, such as radiation (e.g., the fire could reasonably damage equipment in a way such that radiation exposure could be an issue at the location in which the action needs to be taken, requiring the operator to don personnel protection clothing, which takes extra time, but which may not be included in the demonstration); smoke and toxic gas effects which are not likely to be actually simulated in the demonstration (e.g., in a real fire manual for actions needed to be performed near the fire location, although in a separate room, there may be smoke and gas effects that could slow the implementation time for the action); increased noise levels from the fire and the operation of suppression equipment and from personnel shouting instructions; water on the floor possibly delaying personnel movements; obstruction from charged fire hoses; increased heat and humidity resulting from fire-induced loss of heating, ventilation, and air conditioning (heat stress); or too many people getting in each others' way. Again, all these may not actually be simulated, but should be considered as possible, and perhaps even likely, when determining the time it may take to perform the manual action in a real situation.

  –   The demonstration might be limited in its ability to account for (or envelop) all possible fire locations where the actions are needed and for all the different travel paths and distances to where the actions are to be performed. A similar limitation concern is that the current location and activities of needed plant personnel when the fire starts could delay their participation in executing the operator manual actions (e.g., they may typically be at a location that is on the opposite side of the plant relative to a postulated fire location and/or may need to restore certain equipment before being able to participate such as if they are routinely doing maintenance). The intent is not to address temporary/infrequent situations but to account for those that are typical and may impact the timing of the action.

- It may not be possible to execute relevant actions during the demonstration because of normal plant status and/or safety considerations while at power (e.g., operators cannot actually operate the valve using the handwheel, but can only simulate doing so).

(3)   typical and expected variability between individuals and crews leading to variations in operator performance (i.e., human-centered factors), as in the following examples (given the likely experience and training of plant personnel performing the actions, it need not be assumed that these characteristics would lead to major delays in completing the actions, but their potential effects should be considered in the specific fire-related context of the actions being performed, to confirm this assumption):

- physical size and strength differences that may be important for the desired action

- cognitive differences (e.g., memory ability, cognitive style differences)

- different emotional responses to the fire and/or smoke

- different responses to wearing self-contained breathing apparatuses (SCBAs) to accomplish a task, that is, some people may be less comfortable wearing an SCBA face piece (e.g., obscured vision) than other people

- differences in individual sensitivities to "real-time" pressure

- differences in team characteristics and dynamics

The emphasis on adequate time for operator manual actions is consistent, conceptually, with ANSI/ANS-58.8-1994, "American National Standard Time Response Design Criteria for Safety-Related Operator Actions," issued 1994 [Ref. 18], on time response design criteria for safety-related operator actions. That standard established "time response criteria...[that] adopt time intervals...to ensure that adequate safety margins are applied to system and plant design and safety evaluations." The standard recognized that "in actual practice, the operator should be capable of reacting to design-basis events correctly and performing the safety-related operator actions in less time than specified by the criteria in this standard." While this standard was not specifically intended for the actions addressed in this document, the concept embodied in the standard of having adequate time contributes to ensuring the reliability of operator manual actions.

To account for the above variables and uncertainty, it is prudent to determine that adequate time exists when comparing the calculated time available and the time required to perform the action. There are different ways to determine that there is adequate time, as is discussed in Sections 4.2.1 and 4.2.2 of this report. Determining that adequate time is available, accounting for the above variables and uncertainty, along with meeting all the other feasibility and reliability criteria, should provide reasonable assurance that the operator manual actions can reliably be performed under a wide range of conceivable conditions by different plant crews.

3.2.3 Environmental Factors

This criterion addresses the issue that environmental conditions may affect personnel's mental or physical performance of operator manual actions to the extent that, if the actions are not entirely precluded, they could be severely degraded. The expected environmental conditions need to be considered in both the locations where the operator manual actions will be performed and along the access and egress routes. Personnel performance can be degraded, if not precluded, by the inability to reach the location as well as the inability to perform the action in the conditions existing at the location. The environment along the egress route after completion of the operator manual action should also be considered to ensure personnel health and safety throughout.

Environmental factors are those factors that could negatively impact the ability to perform the manual actions, including radiation, lighting, temperature, humidity (caused, for instance, by water from sprinkler operation), smoke, toxic gases, and noise.

That these factors need to be considered follows from such requirements as 10 CFR Part 20, "Standards for Protection Against Radiation," governing radiation exposure in responding to fires [Ref. 20]. As stated in Appendix A to 10 CFR Part 50, "anticipated operational occurrences mean those conditions of normal operation which are expected to occur one or more times during the life of the nuclear power unit...." Fires fall into this category and, therefore, are subject to regulations governing "normal operation," such as 10 CFR 20.1201. Similarly, ANSI/ANS-51.1, "American National Standard Nuclear Safety Criteria for the Design of Stationary Pressurized-Water Reactor Plants" [Ref. 21], and its counterpart, ANSI/ANS-52.1, "American National Standard Nuclear Safety Criteria for the Design of Stationary Boiling Water Reactor Plants" [Ref. 22], consider that a "fire limited to one fire area" (corresponding to "plant condition 2") occurs with a frequency of at least once per year. An event in this frequency range is considered part of "normal operation."

Further, NUREG-0800, Section 9.5.1 [Ref. 2], states that "the strategies for fighting fires in all safety-related areas and areas presenting a hazard to safety-related equipment...should designate...potential radiological and toxic hazards in fire zones; ...ventilation system operation that ensures desired plant air distribution when the ventilation flow is modified for fire containment or smoke clearing operation; ...most favorable direction from which to attack a fire in each area in view of the ventilation direction, access hallways, stairs, and doors that are most likely to be free of fire, and the best station or elevation for fighting the fire." This specific reference is not directly applicable to operator manual actions but applies to firefighting activities that are *not* the subject of this document. However, if, for instance, operator manual actions may need to be performed at locations where potential hazards may exist or specific access paths are recommended to deal with potential environmental concerns, some of this information could be useful for operator manual actions as well. Therefore, the reference is included as a basis.

Emergency lighting is addressed in Section III.J of Appendix R to 10 CFR Part 50 [Ref. 1], or by the plant's approved fire protection program, as well as in NUREG-0800, Section 9.5.1 [Ref. 2], where it is stated that "[l]ighting…[is] vital to safe shutdown and emergency response in the event of a fire."

Studies such as NUREG/CR-5680, "The Impact of Environmental Conditions on Human Performance," Volumes 1 and 2, issued September 1994 [Ref. 23], attest to the impact on human performance of such variables as heat and cold, noise, lighting, and vibration. NUREG-1764, "Guidance for the Review of Changes to Human Actions," issued February 2004 [Ref. 24], cited in NUREG-0800, Section 18.0, Revision 1, February 2004 [Ref. 7], notes that "…[q]ualitative assessment [of the human actions] addresses…the environmental challenges…that could negatively affect task performance…." Experimental studies, such as the ones cited as References 25 and 26, provide further evidence of the effects of heat and cold stresses on the performance of various physical and cognitive human tasks. NUREG-0711, "Human Factors Engineering Program Review Model," Revision 1, issued February 2004 [Ref. 27], also cited in NUREG-0800, Section 18.0 [Ref. 7], states that "[human-system interface] characteristics should support human performance under the full range of environmental conditions, e.g., normal as well as credible extreme conditions…." Accordingly, it needs to be ensured that such habitability issues (including those that may be unique to fire conditions such as additional heat concerns, smoke, toxic gases, effects of ventilation shutdown, the possibility of having to pass through areas and/or manipulate electrical equipment with water on the floor) will not adversely impact the operator manual actions in the locations where the actions are to be taken and along access and egress routes. Experimental studies, such as those cited in References 28 and 29, provide further evidence of the effects of carbon dioxide, for example, on various measures of human performance.

The importance of this criterion is also consistent with current inspection criteria for fire protection manual actions under the environmental considerations' criterion. Current inspections ensure that plant staff has addressed radiation levels per 10 CFR Part 20 [Ref. 20], lighting, temperature and humidity, and fire effects such as smoke and toxic gases. This existing practice and the associated expectations support the need for a criterion that addresses the environment in which the operator manual actions will be performed.

3.2.4 Equipment Functionality and Accessibility

This criterion addresses the need to ensure that the equipment that is necessary to enable implementation of an operator manual action to achieve and maintain postfire hot shutdown is accessible, available, and not damaged or otherwise adversely affected by the fire and its effects (such as heat, smoke, water, combustible products, spurious actuation). Plant SSCs are the means by which hot shutdown conditions are achieved and maintained. Systems and components often require active intervention, through either automatic or manual means, to perform their function. Hence, equipment that may involve operator manual actions to perform its hot shutdown function needs to be identified and verified to be both accessible and functionally available to the extent required to successfully implement the operator manual actions.

Information Notice (IN) 92-18, "Potential for Loss of Remote Shutdown Capability During a Control Room Fire," dated February 28, 1992 [Ref. 30], identifies the type of functionality issue that should be considered. For example, the bypassing of thermal overload protection devices for motor-operated valves (MOVs) (discussed in Regulatory Guide 1.106, "Thermal Overload

Protection for Electric Motors on MOVs," issued March 1977 [Ref. 31]) could jeopardize completion of the safety function or degrade other safety systems due to sustained abnormal circuit currents that can arise from fire-induced "hot shorts." Even if the overload protection devices are not bypassed, hot shorts can cause loss of power to MOVs by tripping the devices. If an operator manual action involves the manual manipulation of a powered MOV, such fire-induced damage (e.g., overtorquing a MOV) could render manipulation physically impossible. Other equipment could also have fire-damage-susceptible parts. Therefore, if equipment (including cabling and power and cooling to support the equipment) that could be affected by the fire or its subsequent effects are planned for use via operator manual actions, the plant staff should verify that the functionality of that equipment will not be adversely affected and the function can be successfully accomplished by manual actions.

Accessibility to these systems and equipment is necessary to enable personnel to perform the operator manual actions on the components. Not only must the personnel be able to find and reach the locations of the components, but they also must be able to perform the required action on the components.

The importance of this criterion is also consistent with current inspection criteria for fire protection manual actions under the accessibility criterion and other related criteria. Current inspections ensure, for instance, that the necessary equipment is available and protected from fire effects. This existing practice and the associated expectations support the need for a criterion that addresses the functionality and accessibility of equipment needed to successfully perform operator manual actions.

3.2.5 Available Indications

In addition to the SSCs needed to directly perform the desired functions, the equipment needs to include diagnostic indications relevant to the desired operator manual actions. These indications, to the extent required by the nature of the operator manual action, may be needed to (1) enable the operators to determine which manual actions are appropriate for the fire scenario, (2) direct the personnel performing the manual actions, and (3) provide feedback to the operators, if not already directly observable, to verify that the manual actions have had their expected results and the manipulated equipment will remain in the desired state. These indications include those necessary to detect and diagnose the location of the fire. As necessary equipment, indications should also meet the functionality and accessibility criterion discussed above.

This indication criterion is consistent with the guidance in Generic Letter (GL) 81-12, "Fire Endurance Test Acceptance Criteria for Fire Barrier Systems Used to Separate Redundant Safe Shutdown Trains Within the Same Fire Area (Supplement 1 to Generic Letter 86-10: Implementation of Fire Protection Requirements" [Ref. 32], regarding manual actions for associated circuit resolution for alternative shutdown (Paragraph III.G.3 of Appendix R to 10 CFR Part 50 [Ref. 1]):

> For circuits of equipment and/or components whose spurious operation would affect the capability to safely shutdown…provide a means to detect spurious operations and then [provide] procedures to defeat the maloperation of equipment (i.e., closure of the block valve if (a power-operated relief valve

(PORV)) spuriously operates, opening of the breakers to remove spurious operation of safety injection).

The adequacy of indications to detect the need for an action (in this example, spurious operations) illustrates the basic concept of needing sufficient indications so that (1), (2), and (3) in the previous paragraph can be performed.

Section IX of Attachment I to IN 84-09, "Lessons Learned From NRC Inspection of Fire Protection Safe Shutdown Systems (10 CFR 50, Appendix R)," dated March 7, 1984 [Ref. 33], lists the minimum monitoring capability, which includes (1) diagnostic instrumentation for shutdown systems, (2) level indication for all tanks used, (3) pressurizer (PWR) or reactor water (BWR) level and pressure, (4) reactor coolant hot-leg temperatures, or core exit thermocouples, and cold-leg temperatures (PWR), (5) steam generator pressure and level (wide range, PWR), (6) source range flux monitor (PWR), (7) suppression pool level and temperature (BWR), and (8) emergency or isolation condenser level (BWR). However, annunciators, indicating lights, pressure gauges, and flow indicators are among the instruments typically not protected under the guidance in IN 84-09, although these instruments may be needed to detect that a maloperation or other trigger for action has occurred. IN 84-09 does not exclude other alternative methods of achieving hot shutdown. Plant staff may employ alternative instrumentation to help achieve hot shutdown (e.g., boron concentration indication).

The importance of providing more indication than recommended in IN 84-09 [Ref. 33] was recognized when the NRC updated its inspection guidance in March 2003 [Ref. 10] for operator manual actions. "Determine whether adequate diagnostic instrumentation,[6] unaffected by the postulated fire, is provided for the operator to detect the specific spurious operation that occurred." Suppose a plant has protected only the instrumentation needed to conform to IN 84-09. If due to lack of circuit protection, the plant staff has to respond to an inappropriate equipment operation (e.g., decreasing pressurizer level), additional diagnostic instrumentation needs to be sufficient for the operator to direct the correct response. For example, the decreasing pressurizer level could be due to spurious closure of an in-line MOV. If so, which one? The plant's fire protection safe shutdown analysis should consider the means to determine the source of the problem, if that is necessary to identify the correct operator action.

The importance of available indication is also covered in such documents as NUREG-1764 [Ref. 24] and NUREG-0711 [Ref. 27], which are cited in NUREG-0800, Section 18.0 [Ref. 7]. NUREG-1764 states that "...a description should be provided for...parameters that indicate that the high-level function is available...operating[, and]...achieving its purpose.... [C]onsider not only the personnel role of initiating manual actions but also responsibilities concerning automatic functions, including monitoring the status of automatic functions to detect system failures...." NUREG-0711 discusses the need to "...provide evidence that the integrated system adequately supports plant personnel in the safe operation of the plant.... The objectives should be to...validate that, for each human function, the design provides adequate alerting, information, control, and feedback capability for human functions to be performed under normal plant evolutions...[and] transients."

---

[6] Defined in GL 86-10, "Implementation of Fire Protection Requirements," dated April 24, 1986 [Ref. 34], as "instrumentation beyond that previously identified in IN 84-09 needed to ensure proper actuation and functioning of safe shutdown and support equipment (e.g., flow rate, pump discharge pressure)."

3.2.6 Communications

In addition to the SSCs needed to directly perform the desired functions, equipment to support communications among personnel may be needed to ensure proper performance of the operator manual actions. For instance, besides the use of face-to-face communication, implementation of an operator manual action may need the use of two-way radios or other electronic or powered forms of communication. In these cases, such equipment may be essential to providing feedback between operators in the MCR and personnel out in the plant, as well as between personnel in different locations of the plant. This communications equipment may then be needed to ensure that any activities requiring coordination among them are clearly understood and correctly accomplished. Further, the unpredictability of fires can force staff to deviate from planned activities (hence, the need for effective, and in some cases, constant communications). Communications permit the performance of sequential operator manual actions (where one set of actions must be completed before another set can be started) and provide verification that procedural steps have been accomplished, especially those that must be conducted at remote locations. Therefore, effective communications equipment, to the extent it is needed, should be readily available and meet the functionality and accessibility criterion covered in Section 3.2.4 above.

The need to emphasize communications equipment is cited, for instance, in NUREG-0800, Section 9.5.1 [Ref. 2], which states "…two-way voice communication…[is] vital to safe shutdown and emergency response in the event of a fire. Suitable…communication devices should be provided…." Further, NUREG-0800, Section 18.0 [Ref. 7], references NUREG-1764 [Ref. 24], NUREG-0711 [Ref. 27], and NUREG-0700, "Human-System Interface Design Review Guidelines," Revision 2, issued May 2002 [Ref. 35], which state that "qualitative assessment [of the human actions] addresses…the level of communication needed to perform the task…. When developing functional requirements for monitoring and control capabilities that may be provided either in the control room or locally in the plant, the following…should be considered: …communication, coordination…workload [and] feedback." Examples cited include "loudspeaker coverage…page stations…personal page devices suitable for high-noise or remote areas…[and] communication capability…for personnel wearing protective clothing [such as] voice communication with masks…." Experimental studies, such as the ones cited in Reference 36, provide further evidence of the effect of respirators on human task performance.

The importance of this criterion is also consistent with current inspection criteria for fire protection manual actions under the communications criterion. Current inspections ensure that the communications capability will be protected from the effects of a postulated fire. This existing practice and the associated expectations support the need for a criterion that addresses communications needed to successfully perform operator manual actions.

3.2.7 Portable Equipment

In addition to the SSCs needed to directly perform the desired functions, the equipment needed to successfully implement the operator manual actions may also include portable equipment relevant to the operator manual actions. Portable equipment, especially unique or special tools (such as keys to open locked areas or manipulate locked controls, flashlights, ladders to reach high places, torque devices to turn valve handwheels, and electrical breaker rackout tools), can be essential to access and manipulate SSCs to successfully accomplish operator manual actions. Hence, to the extent this equipment is needed to successfully implement the operator manual action, this equipment should be readily available and its location should be known and constant. This equipment should be in working order (functional) and access to this equipment should be unimpeded so that it will not delay the operator manual actions and functional.

The importance of this criterion is consistent with current inspection criteria for fire protection manual actions under the special tools criterion ensuring that such equipment is dedicated and available. This existing practice and the associated expectations support the need for a criterion that addresses the use of portable equipment needed to successfully perform operator manual actions.

3.2.8 Personnel Protection Equipment

Besides the SSCs needed to directly perform the desired functions, the equipment needed to successfully implement the operator manual actions may also include personnel protection equipment relevant to the operator manual actions, such as protective clothing, gloves, and SCBAs. Such equipment may need to be worn, for example, to permit access to and egress from locations where the operator manual actions must be performed since the routes could be negatively affected by fire effects, such as smoke that propagate beyond the immediate fire area. Hence, to the extent it is needed to successfully implement the operator manual action, access to this equipment should be unimpeded so that it will not delay the operator manual actions, and this equipment needs to be in working order (e.g., an SCBA must provide a tight seal against any smoke ingress, be in working order when donned, and not malfunction while being used).

NUREG-0800, Section 18.0 [Ref. 7], references NUREG-0700 [Ref. 35], which supports the need to consider this equipment by stating, "[t]he operation of controls should be compatible with the use of protective clothing, if it may be required…. The likelihood of operators requiring protection…is greater outside the control room."

Further, current inspection guidance treats this equipment as subject to the special tools criterion cited previously.

3.2.9 Procedures and Training

This criterion reflects the need for written, maintained plant procedures that cover all the manual actions and the need for each operator who might be required to perform the actions to achieve and maintain hot shutdown to receive training on these manual actions. The role of written plant procedures in the successful performance of operator manual actions is threefold:

(1)   They assist the operators in correctly diagnosing the type of plant event that the fire may trigger (usually in conjunction with indications), thereby permitting the operators to select the appropriate operator manual actions.

(2)   They direct the operators to the appropriate preventive and mitigative manual actions.

(3)   They minimize the potential confusion that can arise from fire-induced conflicting signals, including spurious actuations, thereby minimizing the likelihood of personnel error during the required operator manual actions. Written procedures contain the steps of what needs to be done, and unless it can be argued to be "skill-of-the-craft," they should also contain guidance for how and where it should be done, and what tools or equipment should be used.

Training on these procedures serves three supporting functions—(1) it establishes familiarity with the fire procedures and equipment needed to perform the desired actions, as well as, potential conditions in an actual event, (2) it provides the level of knowledge and understanding necessary for the personnel performing the operator manual actions to be well prepared to handle departures from the expected sequence of events, and (3) it gives personnel the opportunity to practice their response without exposure to adverse conditions, thereby enhancing confidence that they can reliably perform their duties in an actual fire event.

With regard to plant procedures, in general, Appendix B, "Quality Assurance Criteria for Nuclear Power Plants and Fuel Reprocessing Plants," to 10 CFR Part 50 [Ref. 37] requires quality assurance procedures for nuclear power plants:

> Activities affecting quality shall be prescribed by documented instructions [or] procedures...of a type appropriate to the circumstances and shall be accomplished in accordance with these instructions, procedures, or drawings. Instructions [or] procedures...shall include appropriate quantitative or qualitative acceptance criteria for determining that important activities have been satisfactorily accomplished.

Appendix A to Regulatory Guide 1.33, "Quality Assurance Program Requirements (Operations)," issued February 1978 [Ref. 38], on quality assurance programs for power operation describes a method acceptable to the NRC staff for complying with these Appendix B requirements. Appendix A to the regulatory guide identifies the following as typical safety-related activities that should be covered by written procedures—(1) the plant fire protection program (administrative procedures), (2) mode change from plant shutdown to hot standby and operation at hot standby (general plant operating procedures), (3) changing modes of operation for a wide range of safety-related PWR and BWR systems (specific plant operating procedures), and (4) plant fires (procedures for combating emergencies and other significant events). In addition, there should be procedures for abnormal, off-normal, and alarm conditions, with each safety-related annunciator having its own written procedure. In conformance with the above, procedures covering operator manual actions in response to fire should be controlled procedures such as those covering other plant operations. The training portion of this criterion is an extension of the requirement of 10 CFR 50.120, "Training and Qualification of Nuclear Power Plant Personnel" [Ref. 39], that nuclear power plant personnel is trained and qualified. "Each nuclear power plant licensee...shall establish, implement, and maintain a training program derived from a systems approach to training as defined in 10 CFR 55.4 [Operators' Licenses—Definitions, Ref. 40].... The training program must incorporate the instructional requirements necessary to provide qualified personnel to operate and maintain the facility in a safe manner in all modes of operation."

The personnel performing operator manual actions (operators, maintenance staff, electrical technicians) need to undergo training for their individual responsibilities. Existing plant training programs should largely address many of the training issues, such as ensuring instruction is provided by qualified individuals, that it is provided to all personnel who may be required to perform operator manual actions, and that practice sessions are held consistent with the requirements for training on other abnormal procedures for each member of the operating crews that could be involved in diagnosing or performing the actions. This will provide them with experience in performing the operator manual actions.

In addition, as discussed below in the demonstration sections of this report (i.e., Sections 3.2.11 and 4.2.11), there may be some actions that need to be practiced under as realistic conditions as possible, on a regular basis, by all crews (i.e., the actions need to be demonstrated on a regular basis to ensure that they can be performed reliably). For these operator manual actions, actual demonstrations of the actions, under conditions as closely approximating actual fire situations as is reasonable, should become part of the regular training program.

Drills for operator manual actions should address such issues as the effectiveness of alarms in communicating the desired operator action intent for different fires and their locations; operator time response relative to that used in the timing analysis; proper use of portable equipment, including communication devices and personnel protection; each operator's knowledge on his or her role, particularly if the fire results in different role assignments such as interfacing with the Fire Brigade; and conformance with requirements of the plant fire procedures.

Certainly, most of the above characteristics of the procedures and training, and the bases for these characteristics, will already be covered in existing procedure and training programs. Information regarding the implementation with specific regard to postfire operator manual actions is covered in Section 4.2.9 of this report. The focus here should be on ensuring that those unique aspects, due to the fact this is a postfire response, are indeed addressed.

The importance of this criterion is also consistent with current inspection criteria for fire protection manual actions under both the procedures and the training criteria. Under these criteria, inspectors are to ensure that (1) operators do not have to study procedural guidance at length to operate the equipment in the manner intended, and (2) training on the manual actions and the procedure is adequate and current. This existing practice and the associated expectations support the need for a criterion that addresses procedures and related training needed to successfully perform operator manual actions.

3.2.10 Staffing

The intent of the staffing criterion is to ensure that an adequate number of qualified personnel will be available so that hot shutdown conditions can be achieved and maintained in the event of a fire. Credited personnel may be normally on site, or available through the emergency planning staff augmentation system in time to successfully perform the desired action. Further, individuals that might be needed to perform the operator manual actions should not have collateral duties, such as firefighting, security duties, or control room operation, during the evolution of the fire scenario. In other words, enough trained people, without collateral duties during a fire, should be available to ensure that operator manual actions can be completed as needed.

For instance, an operator should not serve as both a Fire Brigade member and be responsible to perform an operator manual action during a fire at the same time (i.e., he/she should not serve both functions concurrently). The operator could serve as a Fire Brigade member on shift provided another operator had his/her manual action responsibility that same shift. The intent is that an individual who could be called upon to perform operator manual actions should not, for example, also be a member of the Fire Brigade for the same fire, or have other duties that would interfere with his/her ability to perform the operator manual action in a timely manner. Therefore, all operating shift staffing levels should include enough trained personnel to perform any operator manual actions that could arise since any fire could occur at any time.

NUREG-0800, Section 18.0 [Ref. 7], cites NUREG-1764 [Ref. 24] and NUREG-0711 [Ref. 27], which in turn provide NRC staff views with regard to staffing. NUREG-1764 states that "[s]taffing levels should be evaluated based on…[r]equired actions…[t]he physical configuration of the work environment…[a]vailability of personnel considering other activities that may be ongoing and for other possible responsibilities outside the control room…." NUREG-0711 states that "[t]he basis for staffing and qualifications should…address…the knowledge, skills, and abilities needed for personnel tasks…availability of personnel…crew coordination concerns that are identified during the development of training." Also, "validate that the shift staffing, assignment of tasks to crew members, and crew coordination (both within the control room as well as between the control room and local control stations and support centers) is acceptable. This should include validation of nominal shift levels, minimal shift levels, and shift turnover…." In addition, "address…personnel response time and workload…the job requirements that result from the sum of all tasks allocated to each individual both inside and outside the control room…the requirements for coordinated activities between individuals…[and] the interaction with auxiliary operators…. [V]alidate that specific personnel tasks can be accomplished within time and performance criteria, with a high degree of operating crew situation awareness, and with acceptable workload levels that provide a balance between a minimum level of vigilance and operator burden…."

The subject of staffing has also been addressed many times before with regard to NRC's intent in this area. For instance, IN 91-77, "Shift Staffing of Nuclear Power Plants," dated November 26, 1991 [Ref. 41], states that "[t]he number of staff on each shift is expected to be sufficient to accomplish all necessary actions to ensure a safe shutdown of the reactor following an event…. Licensees may wish to carefully review actual staffing needs to ensure that sufficient personnel are available to adequately respond to all events. This is especially relevant to the backshift when staffing levels are usually at a minimum…."

This criterion on staffing is similarly addressed in Section III.L of Appendix R to 10 CFR Part 50 [Ref. 1]. It states, "The number of operating shift personnel, exclusive of Fire Brigade members, required to operate the equipment and systems comprising the means to achieve and maintain the hot standby or hot shutdown conditions shall be on site at all times." The NRC contends that, if the Fire Brigade could be expected to perform actions other than those solely involved with firefighting, the potential exists for interfering with either their firefighting activities or the operator manual action, such that successful performance of one or the other, or both, could be impaired. Although it may seem redundant to require an operator, independent of any firefighting responsibility, to perform an action that could simply be performed by a member of the Fire Brigade, one can conceive of situations where this dual responsibility could be a problem. Hence, operators should be independent of the fire brigade duties and even control room duties since operator manual actions take place outside the control room.

Further, the importance of this criterion is consistent with current inspection criteria for fire protection manual actions under the staffing criterion to determine whether adequate qualified personnel are available to perform the operator manual actions. This existing practice and the associated expectations support the need for a criterion that addresses staffing needs relative to performing operator manual actions.

3.2.11 Demonstrations

This criterion provides a degree of overall assurance that the operator manual actions can be performed in the analyzed time available (i.e., the actions are feasible). This criterion provides a "test" (by at least one randomly selected but established crew) that all feasibility and reliability criteria have been and continue to be met. As a result, the desired operator manual actions are shown to be achievable within the constraints, including the analyzed time available, using the minimum staffing levels, with the expected operable equipment, under the expected environmental conditions (to the extent that they can be reasonably simulated), using the procedures and training provided for the manual actions. The plant staff should not rely upon any operator manual action until it has been demonstrated to be consistent with the analysis.

In addition, this criterion and the criterion to show adequate time available to ensure reliability, which includes showing that extra time exists to account for factors that may not or cannot be covered in the demonstration, complement each other. The demonstration serves as a benchmark against which it can be determined how much extra time is needed to cover the potential influences of the factors not modeled in the demonstration that could delay performance, which more directly addresses the reliability concept. As with training, the demonstration provides the crew with practical experience. All elements of the fire scenario, including, but not limited to, diagnosis of the need for the action, the use of equipment and procedures, adequacy of staffing levels, response to indications, should be integrated into the demonstration to the extent possible to develop this benchmark. In this way, any complexities, such as the number of operator manual actions and their dependence upon one another, and the handling of multiple procedures (emergency operating procedures (EOPs), as well as fire plans and procedures) at the same time, are evaluated and identified for appropriate consideration in determining how much extra time is needed.

Failure to show in a demonstration that the operator manual actions can be accomplished in a manner that is consistent with the analysis (i.e., within the time available to ensure that hot shutdown conditions can be achieved and maintained), indicates that the manual actions are not feasible. In such cases, the plant staff could try modifying the actions (e.g., different access/egress routes, redeployment of critical equipment by placing it at the location where the manual action will be performed vs. carrying it to that location, dividing the activities among a greater number of staff), such that a new demonstration satisfies the analysis. Alternatively, the plant staff could conclude that operator manual actions are not feasible and, therefore, opt for another means to assure that hot shutdown can be achieved and maintained (e.g., passive fire protection features, with fire detection and automatic suppression, as appropriate). Another alternative depends on the nature of the calculations and analysis performed to determine how much time is available. If the calculations and analysis made very conservative assumptions (i.e., they produce a nonmechanistic minimum estimate of the time available rather than a more realistic estimate), then it may be possible to make a strong case that more time would actually be available and that the action is therefore feasible. It may be possible to make such an argument by pointing out how the various assumptions would lead to underestimations of the time available and that if more realistic assumptions were made, adequate time would clearly be available. However, if the effects of the conservative assumptions on the calculated time available are difficult to estimate, additional calculations may be required (see Section 4.2.1 of this report).

Plant staff may determine that operator manual actions are feasible after an initial demonstration has been successfully accomplished and it can be shown that the actions can be performed within the time available. Similarly, if it can be shown that the demonstrated time (or estimated time to complete the action based on the demonstration), along with the extra time needed to account for factors not included in the demonstration, can be enveloped by the estimate of the time available, then it can be argued that the actions may also be performed reliably. If this criterion cannot be met, then as noted above, plant staff could take steps to improve performance of the actions, decide that they cannot be performed reliably, or argue that because of conservative assumptions in the calculations, enough time is available to ensure reliability.

Subsequent demonstrations are likely to be needed for the more complex (see below) operator manual actions, but they may not be necessary for all scenarios by all crews. In some cases, the actions may be straightforward enough that they can be covered through regular training and practice on critical aspects of the operator manual action. In other words, subsequent "full-blown" demonstrations, involving as realistically as possible simulation, may not always be necessary, as long as the operating crews that could be involved in diagnosing or performing the actions receive regular training and practice. As discussed earlier, the training and practice should be done at a frequency consistent with that established in existing training programs on abnormal procedures in compliance with 10 CFR 50.120 [Ref. 39]. This will provide them with experience in performing the operator manual actions.

However, for more complex actions, where, for example, significant coordination might be involved or a sequential set of actions must be executed in a specified order, possibly in different locations or involving multiple individuals, subsequent periodic demonstrations should be carried out to ensure that the actions can continue to be performed reliably. Other general examples that might require periodic demonstrations include situations in which the following three complex conditions exist:

(1)   There is a need to decipher numerous indications and alarms.

(2)   There may be ambiguity associated with assessing the situation or in executing the task.

(3)   The activity requires very sensitive and careful manipulations by the operator, particularly in a time sensitive situation.

Since plant staff will rely on the operator manual actions to ensure the safety of the plant, and because NRC inspectors may ask for periodic demonstrations of various operator manual actions, plant staff should identify the actions that require regular, realistic (as possible) demonstrations and ensure that all crews receive adequate participation in those demonstrations.

Subsequent demonstrations provide valuable training and experience for plant personnel and also serve to verify that plant configuration and conditions (e.g., access, egress) have not changed over time such that the manual actions may no longer be accomplished in accordance with the analyzed time available. If plant staff is unable to successfully complete a subsequent demonstration, they should take corrective action to modify the manual action or the conditions contributing to the inability to successfully complete the demonstration. This agrees with the general concept of corrective action as expressed, for instance, in Criterion XVI of Appendix B to 10 CFR Part 50 [Ref. 37], which requires corrective action measures for conditions adverse to quality. If plant staff is unable to complete a successful demonstration, the staff should opt for another means to ensure that hot shutdown can be achieved and maintained (e.g., passive fire protection features, with fire detection and automatic suppression, as appropriate).

The intent of this criterion is to provide assurance that any crew that might be on duty at the time of a fire can reliably perform the operator manual actions, allowing for variability and uncertainties. It should be sufficient that "an established crew" can illustrate the ability to perform the operator manual actions through a demonstration(s) of the relevant actions. In addition, as discussed above, demonstrations of the more complex actions may become part of periodic operator training. To ensure that all crews (including those receiving training but not performing the demonstration during a particular training cycle) could reliably perform the actions, the criterion of "Showing Adequate Time Available to Ensure Reliability" (i.e., extra time is available (see Section 3.2.2 of this report)) is applied to account for variability that exists among crews as well as for likely shortcomings of the demonstration, as discussed previously. In this way, the demonstration by the established crew would support the position that any of the crews could likewise perform the operator manual actions under a wide range of fire situations.

The use of such demonstrations is supported, for instance, by NUREG-1764 [Ref. 24] and NUREG-0711 [Ref. 27], cited in NUREG-0800, Section 18.0 [Ref. 7]. NUREG-1764 states that "...[a] walkthrough of the human actions under realistic conditions should be performed.... The scenario used should include any complicating factors that are expected to affect the crews['] ability to perform the human actions...." NUREG-0711 states that "...an integrated system design (i.e., hardware, software, and personnel elements) is evaluated using performance-based tests.... Plant personnel should perform operational events using a simulator or other suitable representation of the system to determine its adequacy to support safety operations...."

For this criterion, some fire brigade training expectations from Section III.I of Appendix R to 10 CFR Part 50 [Ref. 1] are useful to apply to operator manual actions. Just as fire brigade training includes firefighting practice and fire drills, the personnel performing operator manual actions should participate in a similar program of practice and drills for their actions. However, considering these drills in balance with the uncertainties to be addressed in justifying there is adequate time to perform the operator manual action, it may be that the plant personnel decide to perform these demonstrations under near-simulated fire conditions as a means to lessen the uncertainties to be addressed when justifying there is adequate time. Section III.I of Appendix R states that "Practice sessions shall be held for each shift [crew] to provide them with experience in [performing the operator manual actions] under strenuous conditions encountered [during the fire]. These practice sessions should be provided at least once per year for each [operating crew]... [and] performed in the plant so that the [crew] can practice as a team."

It may be impractical for all the operating crews, unlike the plant fire brigades, to perform the operator manual action demonstrations within a 12-month training cycle. As an alternative, feasibility should be shown through demonstrations (of at least the more complex actions) utilizing an established crew at a frequency that is consistent with an existing training program in compliance with 10 CFR 50.120 [Ref. 39] until all the crews eventually demonstrate the more complex actions. However, since as a minimum, only one crew may actually perform the demonstration within a training cycle, additional considerations are needed to provide reasonable assurance that the operator manual actions can be reliably performed (i.e., repeated successfully by any crew at any time). Also, it is likely that the demonstration cannot simulate all the conditions that might be encountered in an actual situation, making it necessary to extrapolate the demonstration to the expected fire conditions. Again, these concerns are addressed via the criterion to show adequate time available to ensure reliability (i.e., extra time is available to account for such conditions). The more simple operator manual actions would be covered through training and practice sessions as prescribed by the plant's training program.

Additionally, the importance of this criterion is consistent with current inspection criteria for fire protection manual actions under the verification and validation criterion to determine whether the manual actions have been verified and validated by simulating the actions using the current procedure. This existing practice and the associated expectations support the need for a criterion that addresses simulating (i.e., demonstrating) the performance of operator manual actions as a means to verify acceptability.

4. TECHNICAL INFORMATION FOR IMPLEMENTING THE FEASIBILITY AND RELIABILITY CRITERIA

4.1 Overview

This section provides technical information for meeting the feasibility and reliability criteria summarized in Section 3 of this report. As discussed in Section 2.3 regarding the scope of this report, this information focus on the *unique* aspects of the hazard involved (fire) and the potentially unique characteristics of subsequent manual actions during the operators' response.

Collectively, to address both the feasibility and reliability of an action, the first two criteria address the following concerns:

(1)   analysis of the time available to take the desired manual action considering uncertainties in the estimate of the time available

(2)   ensuring there is adequate time considering certain additional uncertainties in the manual action implementation time

Although the approach described below purposely separates (1) and (2) above, it is not intended that plant staff specifically analyze each above concern as separate steps in the analysis process unless the plant staff chooses to address these two criteria in such a two-step fashion. Both criteria can be addressed collectively in one step (and it may be desirable to do so), such as by performing a single-step analysis that can be used to justify both feasibility and reliability, including enveloping possible effects caused by the listed uncertainties.

4.2 Technical Information for Feasibility and Reliability Criteria

4.2.1   Information Regarding the Analysis Showing Adequate Time Available to Perform the Actions to Address Feasibility

For every operator manual action, analyses should show that there is adequate time for the operators to diagnose the need for the actions, travel to action location(s), perform the actions, and confirm the expected response before an undesired consequence occurs, as dictated by the plant staff's determination of the time available to avoid the undesired consequence. An analysis should have the following three elements:

(Element 1)   *An estimate of the time available* to perform the manual action based on a calculation that already exists (e.g., a design-basis calculation) or a new calculation, to ensure that hot shutdown can be achieved and maintained. The estimate of the time available should account for unique fire-related uncertainties that could affect that estimate, such as the following examples:

–   nature of the fire (e.g., whether the fire is a fast energetic fire, failing equipment quickly, or a slow developing fire with little or no equipment failures for some time)

–   reasonable variations in fire detector response times and sensitivities

–   typical variations in air flows that could affect the fire and its growth

–   specific fire initiation location relative to important targets

–   presence (or lack thereof) of temporary but periodic transient combustibles

Note that there are at least two (and perhaps other) ways to account for these uncertainties. One is to perform a conservative analysis (such as using a nonmechanistic assumption that the fire fails everything in a specific location immediately, and yet the detection of the fire triggering operator response is delayed) with a justification that the fire-related uncertainties are enveloped by the conservative analysis. Another is to perform more of a best-estimate analysis such as conducting fire modeling for some fires to estimate equipment damage times, while also accounting for these uncertainties.

(Element 2)___*An estimate of the time to diagnose the need for and implement* the manual action based on input from walkdowns, talkthroughs, judgment, and as substantiated by a demonstration(s). It is preferable that the demonstration replicate, to the extent reasonable, the conditions under which the manual action will have to be performed (see Section 4.2.11 below for information) so that a realistic estimate is made of the total time to diagnose the need for and implement the manual action including (a) the expected diagnosis time (that is, the expected time to confirm the fire, determine its location, and if necessary, determine the need for the action) and (b) the expected execution time (i.e., the expected time to execute the desired action and confirm the desired plant response). The latter might include activities such as the following:

- MCR staff noting the cue(s) of a possible fire
- MCR staff obtaining the correct fire plan and procedures once the fire location is confirmed
- MCR staff informing the plant staff of the fire and calling for fire brigade assembly and actions
- MCR staff alerting and/or communicating with local staff responsible for taking the desired operator manual actions
- MCR staff providing any specific instructions to the responsible local staff for the manual actions
- having the local staff collect any procedures, checking out communications equipment, and obtaining any special tools or personnel protective equipment necessary to perform the actions
- traveling to the necessary locations
- implementing the desired actions noting that some actions may have to be coordinated or done sequentially, that is, cannot start until prior actions are completed and the MCR staff or others are informed, who also may be dealing with the Fire Brigade and handling multiple procedures (EOPs and fire procedures)
- informing the MCR staff and others as necessary that the actions have been successfully completed and the desired effect has been achieved.

(Element 3) *A comparison* of the two times from (1) and (2) (i.e., time available vs. the collective diagnosis and implementation time) *with an accompanying justification/explanation for why there is adequate time available* to complete the action. In the most straightforward case, where it can be shown that the plant staff has calculated a realistic estimate of the time available, it needs only to be shown that the time to diagnose the need for and implement the action (based on the demonstration) is less than the estimated time available. However, if the plant staff has performed an analysis of the time available that produces an estimate closer to the minimum time available (e.g., a nonmechanistic conservative analysis) as opposed to a more realistic estimate, and this results in the calculated available time being less than the time needed to diagnose the need for and implement the action, then the analysis is not so straightforward. In this case, in order to demonstrate feasibility, justification is needed for why the conservative assumptions included in the analysis of the time available (producing an estimate closer to the minimum time available as opposed to a more realistic estimate) are adequate to "make up" the additional time needed to cover the demonstrated time required. In other words, it should be shown (e.g., provide an explanation) that if the conservatism in calculating the available time was removed, the available time would exceed the time required to diagnose the need for and implement the action.

Another alternative for treating cases where the time available is less than the time needed for the action is to modify the analyses of the time available (eliminating the conservatism and obtaining a more realistic estimate) or to modify the temporal requirements of the actions themselves (e.g., use multiple staff instead of one), until adequate time is shown to be available while still meeting the information in this section.

An example of a timeline analysis approach for addressing the action time relative to available time that address issues that need to be considered is presented in Appendix A to this document. The example is meant to provide information for what should be considered and it is not intended to be a criterion. That is, it is not necessary to show that the example was followed; it is simply an illustration that may be useful for analysts.

4.2.2 Information Regarding the Analysis Showing Adequate Time Available to Ensure Reliability

This criterion addresses the reliability of the operator manual actions. While Section 4.2.1 above addresses the three elements necessary to show the feasibility of an operator manual action, an additional element is necessary in the analysis process to provide sufficient evidence that a manual action can reliably be performed:

(Element 4)   A fourth element of the analysis is to *ensure that additional uncertainties in the estimate of the time required to implement the manual action* (listed below) *are accounted for in the analysis* before the final determination is made that adequate time exists for the manual action. Note that, as before, there are at least two ways to account for these additional uncertainties associated with the time required for the manual action. One is to have purposely arrived at a conservative estimate of the time needed to diagnose the need for and implement the manual action (but based generally on the measured demonstration time) with a justification that the *additional uncertainties listed below* are enveloped by the conservative estimate. Another is to specifically account for the additional uncertainties listed below, adding additional time for each applicable uncertainty to the time required for the action as measured from the demonstration.

With respect to the latter option, Appendix B to this report describes an approach that was used to estimate the potential contributions of the various uncertainties to the time required to perform particular actions. However, note that Appendix B describes more than this process. The work described in Appendix B was performed as part of the original rulemaking effort for postfire operator manual actions, which was later discontinued. At that point in time, the potential for including a suggested "time margin"

4-3

for operator manual actions was being considered. That is, providing guidance for how much "extra" time should be shown to be available in order to fully demonstrate the reliability of the manual actions. While the results of that exercise suggested that a factor of 2 would serve as a time margin to reasonably envelop all the uncertainties of concern (i.e., 100 percent of the demonstrated time should be shown to be additionally available), the main reason for its inclusion as an appendix to this document is to illustrate the thought process that was used to address the uncertainties associated with the time to perform the action. Analysts may find the discussion useful in estimating the potential impact of the factors creating the uncertainties so as to ensure that there is adequate extra time, but it is not meant to imply that a factor of 2 should always be shown or that analysts should always use such an approach.

The additional uncertainties described below could increase the demonstrated time required to conduct the operator manual actions. They may originate from human performance issues that may not be possible to cover in the demonstration (i.e., under nonfire conditions) and may not have been otherwise already addressed in the analysis. These uncertainties (also covered in Section 3.2.2 of this report) include the following examples:

(a)   factors that the plant staff likely may be unable to recreate in the demonstration, or in some cases necessarily anticipate for the real fire situation, that could cause further delay in the time it could take to implement the operator manual action under actual fire conditions (i.e., where the demonstration may likely fall short of actual fire situations), as in the following examples:

    –   The operators may need to recover from/respond to difficulties such as problems with instruments or other equipment (e.g., locked doors, a stiff handwheel, or an erratic communication device). Such difficulties can and sometimes do occur and represent a possible uncertainty in how long it will take to perform an action. Having extra time makes it less likely that any such difficulties encountered in a real fire situation will prevent performance of the desired manual action in the time available.

    –   Environmental and other effects might exist that are not easily simulated in the demonstration, such as radiation, for example, the fire could reasonably damage equipment in a way that radiation exposure could be an issue in the location in which the action needs to be taken, causing the need to don personnel protection clothing (which takes extra time), but which may not be included in the demonstration; smoke and toxic gas effects (these are not likely to be actually simulated in the demonstration, but in a real fire where the manual action needs to be taken near the fire location but in a separate room, there may be smoke and gas effects that could slow the implementation time for the action); increased noise levels from the firefighting activities, operation of suppression equipment, or personnel shouting instructions; water on the floor possibly delaying the actions; obstruction from charged fire hoses; heat stress which requires special equipment and precautions; or too many people getting in each others' way. All these may not actually be simulated in a demonstration, but should be considered as possible (and perhaps even likely) when determining the time it may take to perform the manual action in a real situation.

        –   The demonstration might be limited in its ability to account for (or envelop) all possible fire locations where the actions are needed and for all the different travel paths and distances to where the actions are to be performed. A similar limitation concern is that the location or activities of needed plant personnel when the fire starts could delay their participation in executing the operator manual actions (e.g., they may be in a location that is on the opposite side of the plant off the postulated fire location and/or may need to restore certain equipment before being able to participate). The intent is not to address temporary/infrequent situations but to account for those that are typical and may impact the timing of the action.

        –   It may not be possible to execute relevant actions during the demonstration because of normal plant status and/or safety considerations while at power (e.g., operators cannot actually operate the valve using the handwheel, but can only "talk through" doing so).

(b)   factors involving typical and expected variability between individuals and crews leading to variations in operator performance (i.e., human-centered factors), such as the following examples (as noted earlier, given the likely experience and training of plant personnel performing the actions, it need not be assumed that these characteristics would lead to major delays in completing the actions, but their potential effects should be considered in the specific fire-related context of the actions being performed, to confirm this assumption):

        –   physical size and strength differences that may be important for performing the actions

        –   cognitive differences (e.g., memory ability, analytic skills)

        –   different emotional responses to the fire/smoke

        –   different responses to wearing SCBAs to accomplish a task (i.e., some people may be more uncomfortable than others with a mask over their faces, thus affecting action times)

        –   differences in individual sensitivities to "real-time" pressure

        –   differences in team characteristics and dynamics

Only when a comparison similar to that discussed under item 3 in Section 4.2.1 above is done (i.e., a comparison of time available to the collective diagnosis and implementation time), but which accounts for these additional uncertainties and inevitable variability in the time required for the manual action, would this analysis be complete.

4.2.3 Information Regarding Environmental Factors

Environmental conditions encountered by operators while traveling to and from action-related areas, accessing the areas, and performing the operator manual actions should be shown to be consistent with established human factor considerations, including the following:

- Emergency lighting should be provided as required in Section III.J of Appendix R to 10 CFR Part 50 [Ref. 1], or by the plant's approved fire protection program.

- Radiation should not exceed the limits of 10 CFR 20.1201 [Ref. 20].

- Temperature and humidity conditions should not prevent successful performance of the operator manual actions or jeopardize the operator's health and safety. Heat stress analysis may be appropriate for an environment expected to significantly tax the individual in this manner to ensure the desired action can be performed successfully and without harm to the individual.

- Smoke and toxic gases from the fire should not prevent accessing the necessary equipment, hinder successful performance of the operator manual actions or jeopardize the health and safety of the operator. Plant staff should account for expected smoke and toxic gas levels to ensure that they will not affect performance where it is expected such conditions will exist.

If habitable environmental conditions are present when traveling to and from the location(s) where the relevant activities need to take place, as well as at the location(s) itself, the criterion will generally be easily met. However, several other issues also should be considered:

- The donning and wearing of special personnel protective gear such as SCBAs, firefighting turnout gear, gloves, or other protective items to accomplish the operator manual actions in the fire-impacted environment can slow personnel down because of limited visibility or loss of manual dexterity and may hinder their ability to communicate effectively. Reliable communication may be essential if multiple personnel are involved. As discussed in Section 4.2.11 below, if such special gear might be needed in order to successfully complete the operator manual actions, then it is desirable that the gear should be used during the demonstration to substantiate its effectiveness and its impact on the time to complete the actions. While it is possible to perform the desired actions by meeting in "clear" areas to communicate or by going to clear areas where communication devices are located, at a minimum, time delays during the response should be considered. It is desirable that such activities be included in the demonstration if they are going to be used.

- Plant staff should make certain that any special equipment related to environmental conditions, such as protective clothing or flashlights that might be needed for activities in especially dark areas, are readily available and their location constant and known to those who need to use the equipment. Access to this equipment should be unimpeded so that it will not delay the operator manual actions, and this equipment needs to be in working order (functional). These types of activities should be included as part of the demonstration and included in the time to complete the actions.

4.2.4 Information Regarding Equipment Functionality and Accessibility

This criterion addresses the need to ensure that the equipment that is necessary to implement an operator manual action to achieve and maintain postfire hot shutdown is accessible, available, and not damaged or otherwise adversely affected by the fire and its effects. The criterion is meant to ensure that the desired operator manual actions can be successfully performed using that equipment in the manner required by the operator manual action per the applicable procedures and training.

In crediting the functionality of the equipment, the following should be considered:

- Unique fire effects (such as heat, smoke, water, combustible products), and spurious operation that may render the component inoperable by manual or remote manipulation.

- No credit for operator manual actions and the related equipment should be taken involving the use or manipulation of equipment located where it could be exposed to the fire and its effects. If crediting the use of equipment potentially exposed to the fire and its effects is necessary and this should occur only in rare and exceptional circumstances (e.g., using fire-unaffected equipment in an area well after the fire is extinguished), the plant staff should provide justification as to the continued functionality of the component or components for the intended manipulation.[7]

- All the needs of the equipment are to be met for the equipment to be "functionally available." For instance, if the operator manual actions involve the use of a switch and subsequent control signal to a component, the supporting electrical power and signals and associated cabling need to be available. Further, if the equipment's functionality relies on certain support systems (e.g., cooling, ventilation, power, air from a nearby tank) to be manipulated and continue to function (if needed) in the desired manner, those equipment support functions need to also be in working condition and available.

Knowledgeable personnel are to have adequate accessibility to all the necessary equipment and other aids (e.g., diagnostic indications, components to be manipulated, protective clothing, special tools, keys, procedures, communication equipment), and be able to readily locate the equipment and use or otherwise manipulate the equipment in the desired manner per the procedures and training under the anticipated range of fire-related conditions. Considerations in meeting the adequate accessibility criterion should include the following:

- the range of conceivable environmental conditions (see the environmental considerations criterion) under which the actions will be performed, especially radiation and fire-related conditions such as abnormal temperature, radiant energy, and smoke

- physical access or manipulation constraints, especially for locations likely to be congested or where routine operations do not occur or for manipulations not normally performed

---

[7] In doing so, it is preferable that a different exemption request format be used, such as a risk-informed/performance-based exemption via Regulatory Guide 1.174, "An Approach for Using Probabilistic Risk Assessment in Risk-Informed Decisions on Plant-Specific Changes to the Licensing Basis," Revision 1, issued November 2002) [Ref. 42], or one following National Fire Protection Association (NFPA) 805, "Performance-Based Standard for Fire Protection for Light Water Reactor Electric Generating Plants," [Ref. 43], so as to be able to consider the specific fire scenario and the related conditions.

- the possibility that preferred access/egress routes may become inaccessible and alternate routes may need to be used

- the possibility that security doors or similar restraints could be physically or electrically affected by the fire

Consistent with information for equipment functionality, no credit for operator manual actions should generally be taken at locations exposed to the fire and its effects.

An example of the type of functionality issue that should be considered was discussed in Section 3.2.4 of this report with regard to IN 92-18 [Ref. 30]. The information notice focused on MOV functionality due to possible fire damage to control cables. The bypassing of thermal overload protection devices (discussed in Regulatory Guide 1.106 [Ref. 31]) could jeopardize completion of the safety function or degradation of other safety systems due to sustained abnormal circuit currents that can arise from fire-induced "hot shorts." Even if these overload protection devices are not bypassed, hot shorts can cause loss of power to MOVs by tripping the devices. If equipment (including cabling and other support needs such as power and cooling) that could be affected by the fire or its subsequent effects are to be used for operator manual actions, the plant staff should determine that the functionality and performance of that equipment will not be adversely affected so that the function can be successfully achieved by the manual actions.

4.2.5 Information Regarding Available Indications

Diagnostic indicating instrumentation should be among the equipment identified as needed, to the extent it is required to (1) enable the operators to determine which manual actions are appropriate for the fire scenario, (2) tell the personnel how to properly perform the manual actions, and (3) provide feedback to the operators, if not already directly observable, to verify that the manual actions have had their expected results. The available indications should include those indications necessary to detect and diagnose the location of the fire to the extent this information is needed to meet (1) through (3) above. As part of the necessary equipment, indicating instruments should be functional and accessible as discussed earlier, especially in light of the possible harsher than normal conditions in which the indications may need to operate. In addition the following considerations should be addressed:

- The available indications should be all that are needed, either in the MCR or in local areas, to meet, as necessary, (1) through (3) above, including indicators such as annunciators, indicating lights, pressure gauges, flow indicators, and local valve position indicators.

- A review to identify the needed indications should include situations where there are no alarms for potential spurious equipment operations nor any other compelling signal that the equipment status has changed and is detrimental to the safety functions (e.g., a valve shutting that changes the indication of an open lit light to a closed lit light). In such cases, the operator is more likely to miss the change in status and, therefore, not respond to it. To the extent feasible, compensatory measures should be provided. For example, a local operator observes the equipment (part of the staffing requirement), or there are warnings in the procedure to watch for and frequently check specifically identified equipment status relevant to the fire.

The available indications, where necessary, should be sufficiently redundant or diverse such that the operators should be able to detect potential faulty indications as a result of the fire (such as may be caused by failure or spurious operation due to the fire or due to loss of power caused by the fire) and can determine the true plant status by viewing other indications or by getting other independent local operators to verify the suspect indication. Such redundancy and/or diversity considerations should consider where multiple indications could be affected by one spurious fault or failure, such as the loss of a common power supply or a cascading circuit (e.g., a faulty wide range reactor coolant system pressure signal will affect not only the pressure indication but also the subcooling indication because the signal is used to calculate subcooling). Such erroneous indications could be particularly troublesome since, taken together, they may appear normal.

4.2.6 Information Regarding Communications

Adequate communications capability should be illustrated for operator manual actions that need to be coordinated with other plant operations and personnel. Beyond the use of face-to-face interactions, any communications capability necessary to successfully perform the operator manual action (e.g., need to use two-way radios, internal phone system) should be routinely and readily available for all personnel involved in the actions and should be protected from the effects of a postulated fire. It should be noted that the unpredictability of fires can force plant staff to deviate from planned activities (hence, the need for effective, and in some cases, constant communications). In addition, communications permit the performance of sequential operator manual actions (where one action must be completed before another can be started) and provide verification that procedural steps have been accomplished, especially those that must be conducted at remote locations. More information on communications follows and should be considered to the extent applicable for the communication form that is to be used (e.g., face-to-face interaction, use of electronic devices) to implement the operator manual action:

- For the actions of interest, it should be shown that a potential fire will not damage or disable communications equipment if that communications equipment is needed to successfully perform the operator manual action, and that the ability of personnel to successfully use that equipment, given other factors introduced by the fire (e.g., the need to wear protective clothing), will not be adversely affected.

- There should be confirmation that the desired means of communication will work in particularly noisy environments (best done by testing under the noisy condition, if feasible).

- Personnel should have substantial training on activities that involve coordination and communication, including how to clearly state important information. Further, if the means of communication must be set up or otherwise made available, the time to do so should be factored in the time to implement the desired actions.

- As noted in other sections of this document, the plant staff should have shown the ability to communicate while wearing protective gear such as SCBAs, preferably during the demonstration, if that form of communication is likely to be needed to perform the desired operator manual action.

4.2.7 Information Regarding Portable Equipment

Portable equipment may also be needed for some operator manual actions. Portable equipment, especially unique or special tools (such as keys to open locked areas or manipulate locked controls, flashlights, ladders to reach high locations, torque devices to turn valve handwheels, and electrical breaker rackout tools), can be essential to access and manipulate SSCs in accomplishing operator manual actions. Therefore, portable equipment should also be functional and accessible to the extent it is needed to successfully implement the operator manual action. Crediting the use of portable equipment should include the following considerations:

- The portable equipment should be readily available and its location constant and known to those who need to use the equipment. Access to this equipment should be unimpeded so that it will not delay the operator manual actions, and this equipment needs to be in working order (functional).

- The portable equipment should be controlled and it should be routinely verified that the portable equipment is indeed located where it is supposed to be and has not been misplaced or otherwise moved.

- Personnel should be trained to use the special tools and equipment in the planned application.

- If the use of the portable equipment may slow down action implementation, the delay should be considered in the time estimated (and preferably included in the demonstration) to perform the desired actions.

4.2.8 Information Regarding Personnel Protection Equipment

The necessary equipment also includes personnel safety equipment as it is needed to successfully perform the manual actions and prevent harm to personnel. Such equipment could include, for instance, protective clothing, gloves, and SCBAs. Therefore this component also needs to be functional and accessible to the extent it is needed to successfully implement the operator manual action. Considerations for crediting the use of personnel safety equipment should include the following:

- Consideration needs to be given not only to the locations for the operator manual actions, but also to access and egress paths to and from the locations, considering the fire and its effects.

- The personnel safety equipment should be readily available so that its locations are known by those who need to use it, and there will be no delay in obtaining and donning the protective equipment.

- Personnel should be trained to use the protective equipment in the planned application.

- If the use of the protective equipment may slow down the action because of limited visibility, loss of manual dexterity, difficulty in communicating, or other factors, the delay should be considered in the time estimated (and preferably included in the demonstration) to implement the desired actions. Use of SCBAs, including any credit for communication while they are being worn, should only be credited if their capability has been demonstrated by trained personnel. While it may still be possible to perform the desired actions by meeting in clear areas to communicate or by going to clear areas where communication devices are located, at a minimum, time delays during the response should be considered and it is desirable that these activities be included in the demonstration if life support equipment is going to be used.

4.2.9 Information Regarding Procedures and Training

4.2.9.1 Procedures

To help ensure that operator manual actions are performed successfully, procedural guidance for the actions should be readily available, easily accessible, and contained in a maintained and controlled procedure. Operators should generally not rely on having adequate time to locate, review, and implement seldom-used plant procedures to know when and how to operate plant equipment during a fire event. The procedures should accomplish the following:

- Assist the operators (usually in conjunction with indications) in correctly diagnosing the type of plant event that the fire may trigger, thereby permitting them to select the appropriate operator manual actions.

- Direct the operators as to which manual actions are appropriate to place and maintain the plant in a stable, hot shutdown condition for a fire in a given area.

- Minimize the potential confusion that can arise from fire-induced conflicting signals, including spurious actuations, thereby minimizing the likelihood of personnel error when personnel are performing the operator manual actions.

Existing procedural programs will already cover most aspects of ensuring the fire procedures are adequate (e.g., level of detail, ensuring no conflicts). Given the variety of conditions that can occur during a fire, the procedures should also alert personnel to any potentially hazardous conditions that might be generated by fires in particular locations (e.g., expected hazards such as water on the floor caused by firefighting activities in nearby areas). Furthermore, during the development of the procedures, the plant staff should identify any potential "informal rules" that might exist in the plant or biases that might be held by plant personnel about fire conditions and make sure they are addressed in the procedures and during training, if appropriate (e.g., conditions under which personnel should be concerned about interactions between water and electricity).

Finally, there are special considerations for the two general types of operator manual actions in response to fire:

- In the case of preventive actions (i.e., actions that the plant staff expects to take on the basis of the occurrence of a particular fire, without needing further diagnosis, in order to mitigate the potential effects of possible spurious actuations or other fire-related failures so as to ensure that hot shutdown can be reached and maintained), the procedures should be written to cover the possibility that the fire effects occur before the preventive actions are completed. For such cases, the procedures should direct the operators to verify equipment state and position and manually align the equipment as necessary to reach hot shutdown. For these procedures, it is important that operators have a step which directs entry into these preventive actions (e.g., upon verification of a fire in the fire area) to increase the chances that the steps are performed prior to the occurrence of fire damage.

- For reactive actions (that is, actions taken by plant staff during a fire in response to an undesired change in plant status when the staff must diagnose the need for the actions), relevant procedures should clearly describe the indications which prompt initiation of the actions. If redundant cues are available, they should also be addressed in the procedure to aid the operators when the fire causes spurious effects. Crews should be aware that the cues for such actions can, in principle, occur at any time during a fire. If necessary due to timing considerations, such actions may need to be made "continuous action statements" in the fire procedures.

4.2.9.2 Training

Since plant procedures need to include operator manual actions credited to achieve and maintain hot shutdown, each operator that might be required to perform the actions to reach hot shutdown needs to be appropriately trained on those procedures. Training on the fire procedures should accomplish three goals:

(1)   Establish familiarity with the fire procedures, any equipment/controls needing manipulation to perform the desired operator manual actions, and the potential conditions in an actual fire event, including the necessary indications and human-machine interfaces.

(2)   Provide the level of knowledge and understanding necessary to prepare the personnel performing the operator manual actions to handle departures from the expected sequence of events if the need arises.

(3)   Give the personnel the opportunity to practice their response without exposure to adverse conditions, thereby enhancing confidence that they can reliably perform their duties in an actual fire event.

As with the procedures, most of the training needs for performing the desired operator manual actions will already be addressed by an existing training program once the manual actions are included in the training program (e.g., trainer qualifications, use of practice and classroom activities, ensuring training is current).

In addition, there may be some actions that need to be practiced under conditions that are as realistic as possible, on a regular basis, by all crews (i.e., the actions need to be demonstrated on a routine basis to ensure that they can be performed reliably). For these operator manual actions, actual demonstrations of the actions under conditions that closely approximate actual fire situations should be part of the training program (see Section 4.2.11 below for more on demonstrations).

There are several areas in which special (but not unusual) training will be needed to support operators' ability to complete the manual actions in postfire situations:

- All plant personnel who may need to wear protective clothing to perform the actions should receive training in donning the clothing, traveling to the action locations while wearing the protective clothing, and conducting the relevant actions while wearing the protective clothing.

4-12

- Personnel should train on the use of SCBAs and should practice all aspects of the relevant operator manual actions, including communication, while wearing the SCBAs, if they are likely to be required to wear them in an actual fire.

- If communications among personnel are necessary to accomplish the actions, the communications should be part of the training on the actions and should be practiced under conditions that are as realistic as possible for the expected conditions. The personnel should also be well trained in the range of communication equipment that might be necessary. In addition, plant staff should provide guidance and practice on how to best state the relevant information to be understood.

- Along similar lines, if personnel must work as a team to accomplish certain overall goals, such as when having to perform multiple actions in a certain sequence from various locations, they should be given guidance on how to perform effectively as a team to achieve the particular actions and they should practice the actions as a team. Since it is unlikely that "fixed" teams will always be available for specific actions, individuals should have the opportunity to train on the range of activities to achieve the actions.

- The training should include any technical knowledge regarding fires that will be important to ensure adequate operator response to the fire scenario.

With a frequency consistent with that established by the plant staff in compliance with 10 CFR 50.120 [Ref. 39], the plant staff should conduct demonstrations of at least the more complex actions (see Section 4.2.11 of this report for more on this subject) with established crews of operators, showing that the manual actions needed to achieve and maintain the plant in a hot shutdown condition can be accomplished under conditions closely resembling those anticipated in a real fire event.

4.2.10 Information Regarding the Staffing Criterion

To meet the staffing criterion, it is important that the persons involved in performing the operator manual actions be numerous enough and sufficiently qualified to collectively perform the desired actions to achieve and maintain hot shutdown in the event of a fire. Additionally, the following considerations should be addressed:

- Adequate numbers of qualified personnel should be available within the timeframe credited in the analysis for performing the various operator manual actions. Credited personnel may be normally on site, or available through the emergency planning staff augmentation system, as long as the necessary timing of the action(s) can be met.

- Individuals that might be needed to perform the operator manual actions should not have collateral duties, such as firefighting or control room operation, during the evolution of the fire scenario. For instance, an operator should not serve as both a Fire Brigade member and be responsible to perform an operator manual action during a fire at the same time (i.e., he/she should not serve both functions concurrently). The operator could serve as a Fire Brigade member on shift provided another operator had his/her manual action responsibility on that same shift. The intent is that an individual who could be called upon to perform operator manual actions should not, for example, also be a member of the Fire Brigade for the same fire.

Appropriate staffing largely depends on the activities that need to be performed in accordance with the timing and action related analyses discussed earlier. The following should also be considered in evaluating the staffing for the performance of operator manual actions:

- The number of persons should be sufficient to meet the workload assumed in analyses of the time available and the time needed to complete the operator manual action and, as shown under the demonstration criterion, successfully achieve and maintain hot shutdown. Decisions about staffing levels should take into account all operator manual actions that are expected in a particular fire scenario. Since different scenarios may involve different sets of operator manual actions, staffing levels should meet that required for any scenario in terms of the number of staff needed to meet the timing requirements.

- The staff should be trained and qualified in their assigned duties for performing the operator manual actions. This should be performed per the plant's normal training practices and include special considerations given that the desired actions will need to be carried out during a fire (see the procedure and training criterion). Special considerations may include verification of the availability and reliability of instrumentation and equipment, assessing damage to equipment, deenergizing critical equipment to protect it, reenergizing buses, manually manipulating equipment that normally is automatically controlled, implementing fire-specific procedures (including important plant site and offsite notifications), assisting or supporting firefighting activities, and potentially dealing with injuries to plant personnel.

- No single individual should have task assignments nor a task load that results in excessive physical or mental stresses, nor coincident tasks that challenge each person's ability to perform the desired actions in the analyzed times under the range of anticipated conditions. Plant staff should be able to successfully defend their assumptions regarding the ability of the relevant staff to perform under the expected conditions.

4.2.11 Information Regarding How to Perform a Demonstration

This criterion for operator manual actions in response to fire addresses the fact that each action needs to be demonstrated at least once (by one randomly selected but established crew) to show that the feasibility and reliability criteria have been and continue to be met. As a result, the desired operator manual actions should be shown to be accomplishable within the constraints, including the analyzed time available, using the minimum staffing levels, with the expected operable equipment, under the expected environmental conditions (to the extent reasonable), using the procedures and training provided for the manual actions. The plant staff should not rely upon any operator manual action until it has been demonstrated to be consistent with the analysis.

While it is easiest to conceptually imagine each action being individually demonstrated for different fire scenarios, it is acknowledged that some actions and the fire scenario contexts may have characteristics that are very similar (e.g., the actions themselves are similar, timing related to when the actions have to be performed and how long it would take to implement the actions are similar, locations for the actions are not vastly different as to significantly affect travel time to the locations, similar environments exist for the locations for the actions). In such cases, with justification, a demonstration of an action for a given scenario could be argued to bound or otherwise represent other similar actions in similar circumstances. Hence, one demonstration may be sufficient to credit other similar actions under similar situations.

In addition, subsequent demonstrations should be performed for the more complex (see below) operator manual actions, but they may not be necessary for all scenarios by all crews. In some cases, the actions may be straightforward enough that they can be covered through regular training and practice on critical aspects of the operator manual action. In other words, subsequent "full-blown" demonstrations, involving as realistically as possible simulation, may not always be necessary, as long as the operating crews that could be involved in diagnosing or performing the actions receive regular training and practice. As discussed earlier, the training and practice should be done at a frequency consistent with that established by the plant staff for their plant training programs on abnormal procedures in compliance with 10 CFR 50.120 [Ref.39]. This will provide them with experience in performing the operator manual actions.

However, for more complex actions, where, for example, significant coordination might be involved or a sequential set of actions has to be executed in a specified order, possibly in different locations or involving multiple individuals, subsequent periodic demonstrations should be carried out to ensure that the actions can continue to be performed reliably. Other examples that might require periodic demonstrations include situations involving the following complex conditions:

•   There is a need to decipher numerous indications and alarms.

•   There may be ambiguity associated with assessing the situation or in executing the task.

•   The activity requires very sensitive and careful manipulations by the operator, particularly if under time stress.

Since plant staff will rely on the operator manual actions to ensure the safety of the plant, and because NRC inspectors may observe and assess periodic demonstrations of various operator manual actions, plant staff will need to identify the actions that require regular, realistic demonstrations and ensure that all crews receive adequate participation in those demonstrations.

An important purpose of demonstrating the actions and showing that they can be completed in the time available, is to document the feasibility of the actions. However, for the demonstration to be robust, it is desirable that the demonstration be conducted under conditions that are as realistic as possible. Of course, it is clear that, in spite of plant staff's best efforts, there may be conditions that are very difficult, if not impossible, to simulate. This is one of the reasons it is necessary to show that additional time is available beyond that required based on the demonstration (i.e., to provide a way to account for potential shortcomings in the ability to adequately simulate the actual plant conditions during the demonstration). That is, a tradeoff exists between the extent to which the demonstration is realistic, and the uncertainties to be addressed as part of justifying there is adequate time to perform the operator manual action. For instance, more realistic demonstrations translate into less uncertainty with regard to justifying that there is adequate time.

This section provides information on what should be considered and how to ensure that the demonstration is appropriate. One of the first steps in performing a demonstration is to ensure that all relevant aspects of the other feasibility and reliability criteria are met, and that the important characteristics of those criteria are included to the extent possible. In other words, the demonstration should include all aspects that could influence the outcome of the actions, if it is reasonable to do so. Things to consider under each of the criteria are discussed below.

Before proceeding, it should be noted that, to the extent reasonable, the entire fire-induced accident scenario should be simulated for the demonstration, including all the expected MCR activities, if the response to the fire is expected to credit operator manual actions. More details on the nature of the simulation are given below. While it is desirable that any demonstration simulates the fire conditions to the extent reasonable, under all circumstances, the demonstration should be done by taking into consideration the ability to replicate expected fire conditions safely for personnel, and without jeopardizing the safe operation of the plant. All actions associated with detecting and diagnosing the presence of the fire and diagnosing the need for and executing the relevant manual actions should be timed during the demonstration. Obviously, this information will be important in determining whether there will be enough time available to perform the actions.

4.2.11.1 Environment

Once it is determined (per the information in this report) that the relevant actions are possible under the environmental conditions expected to be present in the areas which operators will have to access to complete the actions, as well as in the locations of the actions, it is desirable that those conditions be simulated to the extent reasonable (noting the safety considerations cited above). For example, the following conditions could be simulated in all relevant areas, including areas through which the operators may have to travel:

- the lighting levels expected to be present during the actual fire to the extent feasible and safe (if dangerous to simulate, should be considered in determining how much extra time is needed)

- if the environmental conditions are assumed to involve the use of SCBAs at any time in the scenario, then the donning and wearing of these during those periods

- if protective clothing will be needed at any time, then the donning and wearing of this during those periods

- if SCBAs may be needed, then any communications anticipated during those periods when the SCBAs are worn (assumes personnel who use SCBAs receive training and are qualified in their use)

- the noise levels expected to be present during the fire scenario, if feasible

4.2.11.2 Equipment Functionality and Accessibility

Accessibility to the relevant systems and equipment is necessary to enable the personnel to perform the operator manual actions. To the extent possible, the personnel participating in the demonstration should carry out the actions if the actions can be done without affecting the safety of the plant (e.g., manually open a valve with the handwheel). If the demands of the task and the time to complete the actions must be based on the judgments of plant personnel, then a process should be used to help ensure that the estimates are reasonable (e.g., get multiple independent judgments). A preferred approach is to obtain estimates of the time to execute specific actions when safety is not a concern (e.g., during shutdown or when the system is out of service for some reason).

In addition, if the plant history indicates that certain equipment tends to have persistent types of problems (e.g., a tendency for valve hand wheels to be stiff), then those conditions should be assumed for the demonstration and not preconditioned solely for the demonstration.

4.2.11.3 Available Indications and Main Control Room Response

In conducting the demonstration the actual effects of the fire conditions should be simulated, to the extent possible, in the plant training simulator and the operators should diagnose the need for the relevant actions based on the expected pattern of indications. In other words, the presence of the cues needed to detect the fire should be simulated, and the crew should have to respond accordingly. The MCR response to the scenario should be the same as during an actual fire. The MCR crew should enter the relevant procedures based on the expected indications and take the necessary steps to respond to the fire and reach hot shutdown. The parameters indicating the need for the operator manual actions in response to the fire should also be simulated, and the crew should have to summon the staff necessary for the manual actions, retrieve the relevant procedures, provide the necessary guidance, and interact with the individuals as necessary while they complete the actions for the demonstration. In addition, the personnel executing the actions should have to check relevant indications of successful completion of the actions and verify completion. These indications should be accurately simulated to the extent possible.

All aspects of the scenario associated with diagnosis and the execution of the actions should be timed. This will provide information relevant to determining the time to diagnose the need for the actions and the time needed to implement the actions. If any aspects of the scenario cannot be simulated, their potential impact on the time should be estimated.

4.2.11.4 Communications

The communications necessary to complete the operator manual actions should be part of the demonstration. This should include communications necessary from the detection of the fire through completion of the actions. Examples of conditions that should be included in the demonstration include the following:

- If it cannot always be assumed that the personnel expected to perform the actions will be in the control room at the time they will be needed, then consideration for where the personnel might be with respect to being able to communicate with the control room should be included in the demonstration. If personnel might be in areas where someone would have to be sent to get them, then this activity should be simulated.

- If personnel must be able to communicate with each other and with the control room, then those communications should be part of the demonstration.

4.2.11.5 Portable Equipment

Any portable equipment that will be needed to conduct the operator manual actions during a real fire should also be accessed and used to the extent reasonable during the demonstration. Portable equipment includes unique or special tools, such as keys to open locked areas or manipulate locked controls, flashlights, ladders to reach high places, torque devices to turn valve handwheels, and electrical breaker rackout tools. Such equipment should be located where it would be expected to be located during a real fire. The equipment should not be gathered together and made easily accessible just for purposes of the demonstration (i.e., no preconditioning).

4.2.11.6 Personnel Protection Equipment

Similar to the portable equipment noted above, any personnel protection equipment such as protective clothing, gloves, and SCBAs should be located, accessed, and donned as during an actual fire.

4.2.11.7 Procedures and Training

All activities associated with the use of procedures should be addressed in the demonstration, including the following:

- detection of the entry conditions for the procedures

- retrieval of the procedures

- the potential need for multiple copies

- usability of the procedures under the expected condition (e.g., lighting levels, a place to put them during their execution if they must be closely followed)

In addition, while the selection of a crew for the demonstration should be random, it should be ensured that there has been no preconditioning such as limiting the selection to only those crews most recently trained.

4.2.11.8 Staffing

Staff who will have duties associated with successful completion of the actions (including diagnosis and execution of the actions) should participate. Staffing issues such as the following should be considered in the demonstration:

- If personnel will have to be summoned from outside the MCR, how long it will take them to get to the control room should be assessed as part of the demonstration considering the likely starting locations for the personnel based on where these persons are typically located. Plant staff should consider the potential for the personnel to be in remote locations from which it is difficult to egress and that the personnel may have to complete some actions before they can leave an area if this is the typical situation for some staff members. It would then be preferred that these considerations be included in the demonstration.

- If the actions will involve multiple staff in certain sequences, then these activities, their coordination, and their associated communication aspects should be included.

- If the MCR crew is likely to be directing and coordinating multiple teams involved in executing manual actions, these activities should be simulated. Furthermore, if the individuals in the MCR coordinating these activities will have other significant responsibilities, those responsibilities should also be simulated.

4.2.11.9 Other Aspects Important to the Demonstration

There are several other important issues or aspects that plant staff should consider in conducting an appropriate demonstration:

- If the operator manual actions being examined are preventive actions and it is reasonable that the fire could negatively affect the relevant equipment before the preventive actions are completed, then the participating personnel may need to verify equipment state and position and manually align the equipment as necessary (i.e., take an additional reactive action to restore the equipment to the desired state). Thus, the implementation time for the actions should include the time it would take plant personnel to complete the actions necessary to manually place the affected equipment in its desired state.

- If the operator manual actions being examined are reactive actions, then the plant staff should be aware that the cues for the need for such actions and the associated effects could, at least in principle, occur at any time after the fire starts. Thus, the effects could occur early, during the diagnosis stage of the scenario, or sometime after that. For purposes of the demonstration, plant staff should estimate when the worst-case timing for the occurrence of the spurious fire effects on the relevant equipment would be with respect to the level of activity in the MCR and the plant in general. That is, the equipment should be assumed to be affected at that time which leaves the shortest available (allowable) time to perform the desired actions. This could be based on a nonmechanistic conservative assumption or, for instance, based on some level of best-estimate fire modeling such as accounting for the location of the fire source(s) and fire size(s) relative to relevant equipment so as to determine when the equipment could be damaged. Other factors that might be considered in determining when to assume that the equipment is affected by the fire might include potential interactions with and effects on other equipment.

- If the fire or other factors could affect where personnel have to travel (e.g., what routes they have to take) and where they have to enter and exit various rooms, then this should be considered in the modeling for the demonstration and determining the travel time.

- If the conditions that could be generated by the fire have the potential to vary significantly, this should be accounted for when deciding how to model the scenario(s) for purposes of the demonstration.

- If smoke could significantly affect visibility, the action should generally not be credited.

In general, plant staff should strive to make the demonstrations as realistic as possible and make conservative assumptions as necessary. If this is done and the above information is followed, then the resulting demonstrations, in conjunction with determining adequate time considering certain uncertainties, should achieve the goal of crediting only feasible and reliable operator manual actions.

5. REFERENCES

1. *U.S. Code of Federal Regulations,* "Fire Protection Program for Nuclear Power Facilities Operating Prior to January 1, 1979," Paragraph III.G.2 and Sections III.I, III.J, and III.L of Appendix R to Part 50, Title 10, "Energy."

2. U.S. Nuclear Regulatory Commission "Standard Review Plan for the Review of Safety Analysis Reports for Nuclear Power Plants," Section 9.5.1, "Fire Protection Program," Rev. 4, BTP CMEB9.5-1, "Guidelines for Fire Protection for Nuclear Power Plants," NUREG-0800, October 2003.

3. U.S. Nuclear Regulatory Commission, "Rulemaking Plan on Post-Fire Operator Manual Actions," SECY-03-0100, June 2003. (ADAMS Accession No. ML023180599)

4. U.S. Nuclear Regulatory Commission, "Regulatory Expectations with Appendix R, Paragraph III.G.2, Operator Manual Actions," Regulatory Issue Summary 2006-10, 2006.

5. *U.S. Code of Federal Regulations,* "Fire Protection," Paragraph 50.48, Title 10, "Energy."

6. U.S. Nuclear Regulatory Commission, "Fire Protection Program for Nuclear Power Plants," Regulatory Guide 1.189, Rev. 1, March 2007.

7. U.S. Nuclear Regulatory Commission, "Standard Review Plan for the Review of Safety Analysis Reports for Nuclear Power Plants," Section 18.0, "Human Factors Engineering," NUREG-0800, Rev. 1, February 2004.

8. *U.S. Code of Federal Regulations,* General Design Criterion 3, "Fire Protection," Appendix A, "General Design Criteria for Nuclear Power Plants," to Part 50, Title 10, "Energy."

9. *U.S. Code of Federal Regulations,* "Specific Exemptions," Paragraph 50.12, Title 10, "Energy."

10. U.S. Nuclear Regulatory Commission, "Fire Protection (Triennial)," NRC Inspection Procedure, Attachment 71111.05T, Enclosure 2, "Inspection Criteria for Fire Protection Manual Actions," March 2003 and April 2006.

11. U.S. Nuclear Regulatory Commission, "Perspectives Gained From the Individual Plant Examination of External Events (IPEEE) Program," NUREG-1742, Volumes 1 and 2, April 2002.

12. Nowlen, S.P., M. Kazarians, N. Siu, and H.W. Woods, "Fire Risk Insights from Nuclear Power Plant Fire Incidents," *Fire and Safety 2001*, Elsevier Publishing Co., London, United Kingdom, February 2001.

13. Cooper, S.E., D.C. Bley, J.A., Forester, A.M. Kolaczkowski, A. Ramey-Smith, C. Thompson, D.W. Whitehead, and J. Wreathall, "Evaluation of Human Performance Issues for Fire Risk," *Proceedings of the International Topical Meeting on Probabilistic Safety Assessment PSA '99: Risk-Informed, and Performance-Based Regulation in the New Millennium, August 22–26, 1999, Washington, DC,* M. Modarres, ed., pp. 964–969, American Nuclear Society, La Grange Park, Illinois, 1999.

14. Forester, J.A., S.E. Cooper, D.C. Bley, A.M. Kolaczkowski, N. Siu, E. Thornsbury, H.W. Woods, and J. Wreathall, "Potential Improvements in Human Reliability Analysis for Fire Risk Assessments," *Proceedings of the OECD/NEA/CSNI Workshop on Building the New HRA: Errors of Commission from Research to Application, May 7–9, 2001, Rockville, Maryland, USA.*

15. Electric Power Research Institute and U.S. Nuclear Regulatory Commission, "EPRI/NRC-RES Fire PRA Methodology for Nuclear Power Facilities," Volume 2, "Detailed Methodology," EPRI TR-1011989 and NUREG/CR-6850, September 2005.

16. Gertman, D.I., H.S. Blackman, J. Byers, L. Haney, C. Smith, and J. Marble, "The SPAR-H Method," NUREG/CR-6883, U.S. Nuclear Regulatory Commission, Washington, DC, August 2005.

17. U.S. Nuclear Regulatory Commission, "Technical Basis and Implementation Guidelines for A Technique for Human Event Analysis (ATHEANA)," NUREG-1624, Rev.1, May 2000.

18. American Nuclear Society, "American National Standard Time Response Design Criteria for Safety-Related Operator Actions," ANSI/ANS Standard 58.8-1994, La Grange Park, Illinois.

19. U.S. Nuclear Regulatory Commission, "Crediting of Operator Actions In Place of Automatic Actions and Modifications of Operator Actions, Including Response Times," Information Notice 97-78, October 23, 1997.

20. *U.S. Code of Federal Regulations,* "Occupational Dose Limits for Adults," Paragraph 20.1201, Title 10, "Energy."

21. American Nuclear Society, "American National Standard Nuclear Safety Criteria for the Design of Stationary Pressurized-Water Reactor Plants," ANSI/ANS-51.1-1983, R1986, La Grange Park, Illinois.

22. American Nuclear Society, "American National Standard Nuclear Safety Criteria for the Design of Stationary Boiling-Water Reactor Plants," ANSI/ANS-52.1-1983, R1988, La Grange Park, Illinois.

23. U.S. Nuclear Regulatory Commission, "The Impact of Environmental Conditions on Human Performance," NUREG/CR-5680, Volumes 1 and 2, September 1994.

24. U.S. Nuclear Regulatory Commission, "Guidance for the Review of Changes to Human Actions," NUREG-1764, February 2004.

25. Vasmatzidis, R.E. Schlegel, and P.A. Hancock, "An Investigation of Heat Stress Effects on Time-Sharing Performance," *Ergonomics*, Vol. 45, No. 3, pp. 218–239, 2002.

26. Pilcher, J.J., E. Nadler, and C. Busch, "Effects of Hot and Cold Temperature Exposure on Performance: A Meta-Analytic Review," *Ergonomics*, Vol. 45, No. 10, pp. 682–698, 2002.

27. U.S. Nuclear Regulatory Commission, "Human Factors Engineering Program Review Model," NUREG-0711, Rev. 1, February 2004.

28. Sheehy, J.B., E. Kamon, and D. Kiser, "Effects of Carbon Dioxide Inhalation on Psychomotor and Mental Performance During Exercise and Recovery," *Human Factors*, Vol. 24, No. 5, pp. 581–588, 1982.

29. Sun, M., C. Sun, and Y. Yang, "Effect of Low-concentration Co_2 on Stereoacuity and Energy Expenditure," *Aviation, Space, and Environmental Medicine*, Vol. 67, No. 1, January 1996.

30. U.S. Nuclear Regulatory Commission, "Potential for Loss of Remote Shutdown Capability During a Control Room Fire," Information Notice 92-18, February 28, 1992.

31. U.S. Nuclear Regulatory Commission, "Thermal Overload Protection for Electric Motors on Motor Operated Valves," Regulatory Guide 1.106, March 1977.

32. U.S. Nuclear Regulatory Commission, "Fire Endurance Test Acceptance Criteria for Fire Barrier Systems Used to Separate Redundant Safe Shutdown Trains Within the Same Fire Area (Supplement 1 to Generic Letter 86-10: Implementation of Fire Protection Requirements)," Generic Letter 81-12.

33. U.S. Nuclear Regulatory Commission, "Lessons Learned From NRC Inspections of Fire Protection Safe Shutdown Systems (10 CFR 50, Appendix R)," Information Notice 84-09, Section IX of Attachment I, March 7, 1984.

34. U.S. Nuclear Regulatory Commission, "Implementation of Fire Protection Requirements," Enclosure 2, "Appendix R Questions and Answers," Generic Letter 86-10, April 24, 1986.

35. U.S. Nuclear Regulatory Commission, "Human-System Interface Design Review Guidelines," NUREG-0700, Rev. 2, May 2002.

36. Zimmerman, N.J., C. Eberts, G. Salvendy, and G. McCabe, "Effects of Respirators on Performance of Physical, Psychomotor, and Cognitive Tasks," *Ergonomics*, Vol. 34, No. 3, pp. 321–334, 1991.

37. *U.S. Code of Federal Regulations,* "Quality Assurance Criteria for Nuclear Power Plants and Fuel Reprocessing Plants," Appendix B to Part 50, Title 10, "Energy."

38. U.S. Nuclear Regulatory Commission, "Quality Assurance Program Requirements (Operation)," Regulatory Guide 1.33, Rev. 2, Appendix A, February 1978.

39. *U.S. Code of Federal Regulations,* "Training and Qualification of Nuclear Power Plant Personnel," Paragraph 50.120, Title 10, "Energy."

40. *U.S. Code of Federal Regulations,* "Conditions of Licenses," Paragraph 50.54, Title 10, "Energy."

41. U.S. Nuclear Regulatory Commission, "Shift Staffing at Nuclear Power Plants," Information Notice 91-77, November 26, 1991.

42. U.S. Nuclear Regulatory Commission, "An Approach for Using Probabilistic Risk Assessment in Risk-Informed Decisions on Plant-Specific Changes to the Licensing Basis," Regulatory Guide 1.174, Rev. 1, November 2002.

43. National Fire Protection Association, "Performance-Based Standard for Fire Protection for Light Water Reactor Electric Generating Plants," NFPA 805, Quincy, Massachusetts, 2006.

APPENDIX A

GUIDELINES FOR USING TIMELINES TO DEMONSTRATE SUFFICIENT TIME TO PERFORM THE ACTIONS

APPENDIX A
GUIDELINES FOR USING TIMELINES TO DEMONSTRATE SUFFICIENT TIME TO PERFORM THE ACTIONS

This appendix provides information for using timelines to investigate and illustrate that sufficient time exists to perform the postfire operator manual actions.[1] It is an additional tool to support the assessment of operator manual actions to be used by analysts if desired. The appendix addresses issues involved in making such an assessment, such as the impact of multiple, serial, or parallel actions and differing considerations for preventive and reactive actions. In conjunction with the information in the body of this report, the goal is to illustrate that there is adequate time available to perform all relevant actions and account for additional uncertainties that might not be covered by the demonstration of the action. The approach includes the use of timelines to show that there is sufficient time to diagnose the need for the actions, travel to action locations, perform the actions, and confirm the expected response. The timeline approach should have the following elements, as illustrated in Figure A-1:

(1)   The time of fire detection (T_0), which begins the timeline and represents the first indication that a fire may exist, or at least suspect that a fire has begun. Detection may be via alarms, indicators, an observation from a roving operator, or other means.

(2)   An expected diagnosis time (that is, the expected time to confirm the fire and determine its location). This time is obtained from the demonstration (see the demonstration criterion discussion later) and T_1, the end of the diagnosis time, is to be marked on the timeline.

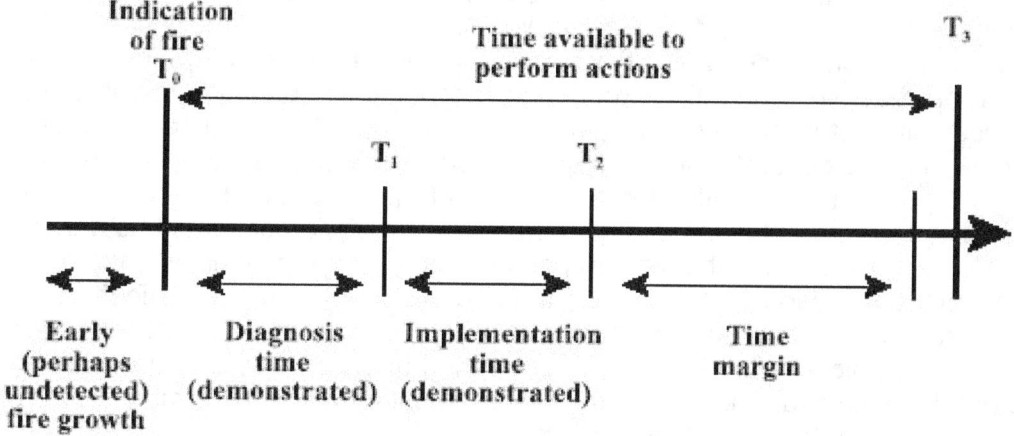

Figure A-1. A timeline

---

(3)   An implementation time that is the expected time to implement the desired action or actions. This time is obtained from the demonstration (see the demonstration criterion) and includes such activities as MCR staff obtaining the correct fire plan and procedures once the fire location is confirmed; informing the plant staff of the fire; calling for fire brigade assembly and actions; calling for and/or communicating with local staff responsible for taking the desired local manual actions; providing instructions to the responsible local staff for the manual actions; having the local staff collect any procedures, checking out communications equipment, and obtaining any special tools or clothing necessary to perform the actions; traveling to the necessary locations; implementing the desired actions (some actions may have to be done sequentially (i.e., cannot start until prior actions are completed)) and communicating with the MCR staff or others as necessary, who in turn may be simultaneously dealing with the Fire Brigade, handling, for example, multiple procedures (emergency operating procedures and fire procedures); and telling the MCR staff and others as necessary that the actions have been completed and the expected effect has been achieved. The implementation time ends at T_2, as shown in the figure. Hence, the total time to be obtained from the demonstration begins at T_0 and ends at T_2.

Note that after the initial diagnosis time, subsequent actions may or may not include subsequent diagnosis times. For instance, in the case of performing proceduralized preventive actions, no other diagnosis time may be needed for some actions. Alternatively, if the desired action is a reactive action in the sense that it is taken only after diagnosis of an undesired equipment status (e.g., loss of feedwater after a valve spuriously closes), then that diagnosis time needs to be included (e.g., deciding what action to take and by whom) as illustrated in Figure A-2. The time available (T_3, the time available to ensure hot shutdown can be achieved and maintained) to complete these reactive actions will need to be measured from the worst-case point at which the equipment could be affected. In other words, since spurious effects caused by the fire could, in principle, occur at any time, analysts would need to determine the point at which the least amount of time would be available to complete the reactive action and successfully restore the availability of the equipment. This could be based on a nonmechanistic conservative assumption or, for instance, based on some level of fire modeling such as accounting for the location of the fire source(s) and fire size(s) relative to relevant equipment so as to determine when the equipment could be damaged. As illustrated in Figure A-2, the starting point for the reactive actions will not necessarily be tied to the time associated with detecting and diagnosing the fire (T_1 in the figures). The symptoms for the reactive actions will occur whenever the fire affects the relevant equipment, which could be before T_1 is reached or anytime after that point. Thus, to repeat, the time available (T_3) for the reactive actions will be determined assuming the worst-case point for the spurious effects.

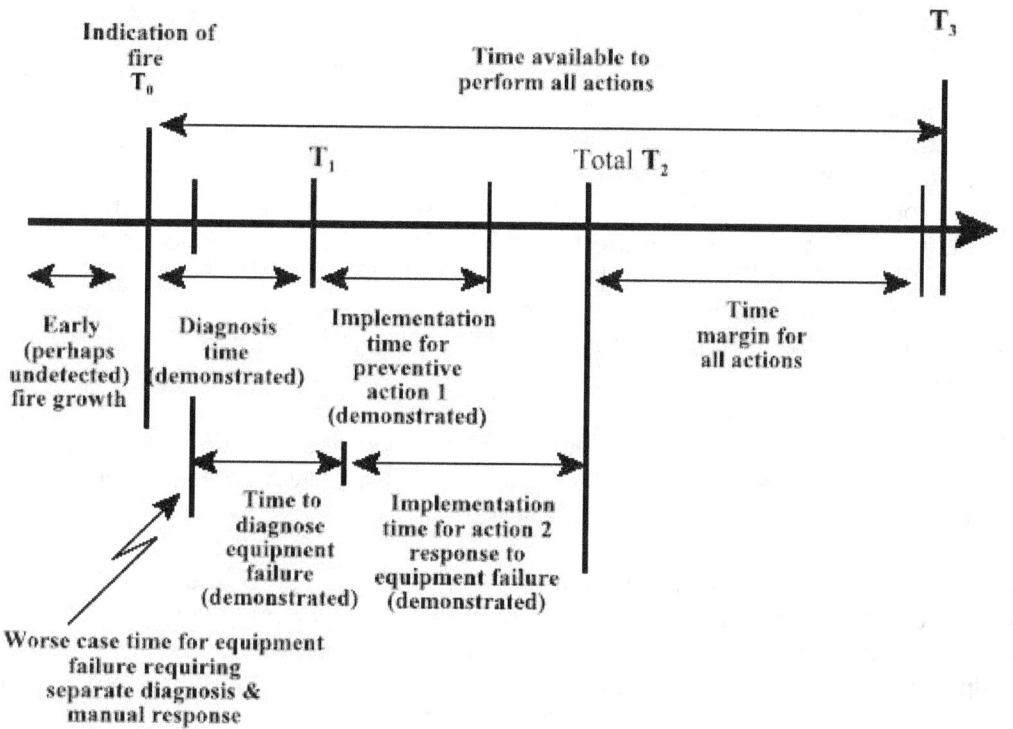

Figure A-2. Initial fire detection and multiple action (one action dependent on a separate diagnosis of an undesired equipment faulure) with a single overall time margin and T_3

Another consideration is relevant to the case of preventive actions. If it is reasonably possible that the fire could negatively affect the relevant equipment before the preventive actions are completed, then the implementation time (T_2) should also include the time it will take plant personnel to take the reactive actions necessary to manually place the affected equipment in the desired state. In other words, when reasonable, analysts should assume that the time to complete the desired action may have to include additional time to take further reactive actions to restore the equipment to the desired state if the fire could negatively affect the equipment before the preventive actions (that are meant to prevent the undesired state of the equipment) can be completed. This issue is addressed further in the information for performing the demonstration.

(4)   Extra time or an added "time margin" to account for uncertainties in estimating the time to complete the action. (A method for determining the "time margin" is explained in Appendix B.)

(5)   The time available for performing the actions to ensure hot shutdown can be achieved and maintained (T_3). T_2 plus the time margin should be less than or equal to T_3.

The relationship between having enough time and the associated demonstration is discussed in detail later. In calculating T_3, it should be shown that the available time is the most conservative (in this case, generally the shortest or minimum) time, considering the fire, its location and anticipated growth rate, the fire effects, and expected plant and operator responses to the fire effects, including thermal-hydraulic calculations as necessary. To determine the most conservative T_3, which in this case is the minimum time available, since overestimating the time available could lead analysts to incorrectly conclude that the actions are feasible and reliable, the analyst needs to consider what failures (including spurious events) may occur and when they may occur. For example, if it is most conservative to assume the equipment failure occurs at the quickest possible time for the fire being analyzed (which may be even before any preventive actions could be taken for the fire, requiring subsequent response-type actions instead), then T_3 should be based on that assumption. For instance, loss of the feedwater function is generally more severe if it happens early in the scenario than if it happens later after a period of successful decay heat removal. If instead it is most conservative to assume the equipment failure occurs at some later time in the scenario, that time should be assumed in deriving T_3 (e.g., if failure of service water to a diesel after the diesel has been running and loaded is more severe than before the diesel is demanded because the diesel could fail in 3 minutes without cooling, so that the operator would likely prevent diesel operation, thereby "saving" it for future use if service water is restored).

As shown in Figure A-3, when developing any timeline showing multiple actions, any interdependence among actions needs to be accounted for, such as when actions by one operator cannot start before another action or actions are completed by another operator, or when multiple actions are to be performed by a single operator who must travel to multiple locations to perform his/her assigned actions in a sequential manner.

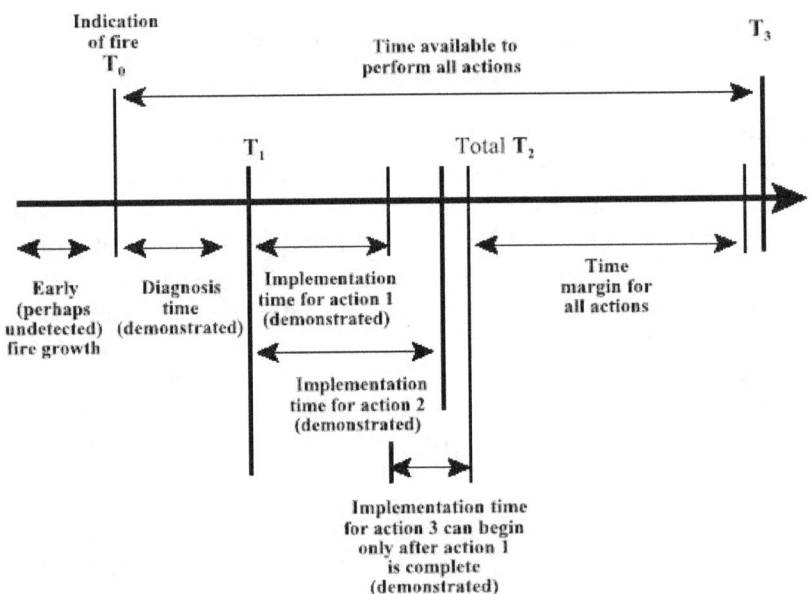

Figure A-3. Initial fire detection and multiple actions
(one action dependent on completion of a prior action) with a single overall time
margin and T_3

A-4

Depending on the desired actions, one overall time margin or multiple time margins and T_3 times (as illustrated in Figure A-4) may be necessary or appropriate to show that individual actions are performed before their specific analyzed T_3 times and that the collective set of actions to fully achieve and maintain hot shutdown are successfully performed considering the fire and its effects. Also, the analysts may wish to use a "most conservative" timeline for a range of fires, locations, and effects (in which case the timeline must envelop the needs of all the fires) or to develop separate timelines for different fire locations or even different fires in the same location.

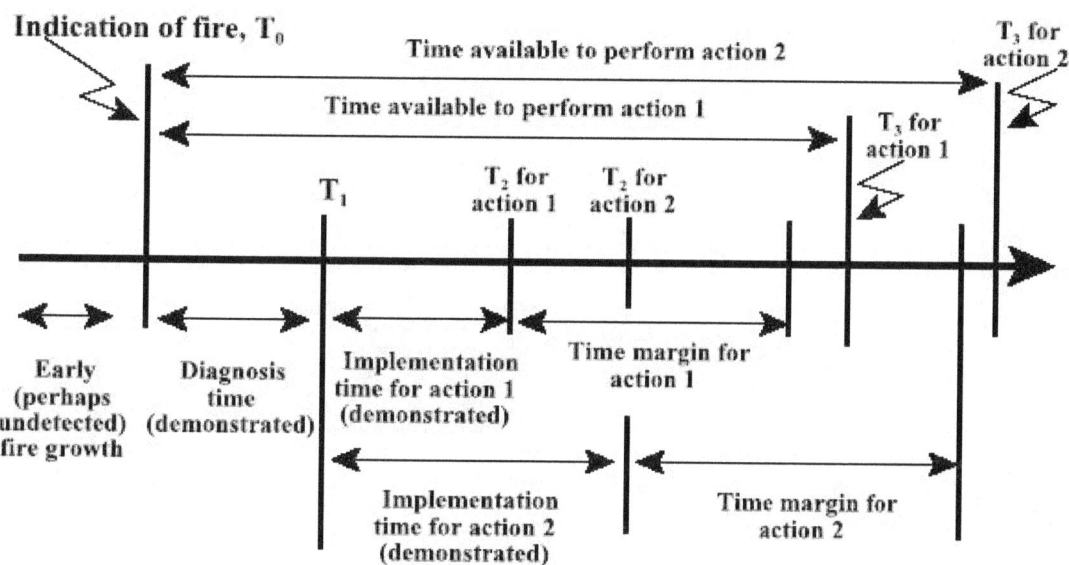

Figure A-4. Initial fire detection and multiple actions illustrating the application of multiple time margins and T_3s

Key inputs and assumptions associated with the timeline should be evident in the analysis documentation.

APPENDIX B

SUMMARY OF EXPERT OPINION ELICITATION
TO DETERMINE TIME MARGINS
FOR OPERATOR MANUAL ACTIONS IN RESPONSE TO FIRE
(April 1–2 and May 4–5, 2004)

APPENDIX B
SUMMARY OF EXPERT OPINION ELICITATION TO DETERMINE TIME MARGINS FOR OPERATOR MANUAL ACTIONS IN RESPONSE TO FIRE
(April 1–2 and May 4–5, 2004)

This appendix documents a process that was used during the initial efforts of the U.S. Nuclear Regulatory Commission's (NRC's) fire operator manual action proposed rulemaking and the development of an accompanying regulatory guide. Since the rulemaking effort was discontinued, this information is included as an appendix to this report to illustrate the thought processes that were used to address the uncertainties associated with the time to perform an operator manual action in response to fire. As such, this appendix describes an expert elicitation that was used to estimate how much additional time analysts might want to include to account for uncertainties that might not be covered by a demonstration of a given action. Analysts may find the discussion useful to their efforts associated with estimating the potential impact of the factors creating the uncertainties to ensure that there is adequate extra time, but it is not meant to imply that a factor of 2 extra time should always be shown or that analysts should always use such an approach in their analysis.

B.1 Introduction

This appendix summarizes the results from two expert opinion elicitation meetings held at NRC headquarters in Rockville, Maryland, to develop quantitative criteria to support the operator manual actions rulemaking [Ref. 1]. The NRC was developing these criteria to ensure that *feasible* operator manual actions could also be accomplished *reliably*, even when considering different levels of complexity, number of actions, etc.[1] Based on an initial meeting held on January 22–23, 2004, among NRC staff and contractors to discuss potential options for quantitative criteria, it was agreed that the use of "time margins" was appropriate as a surrogate for ensuring a high reliability in the credited local operator manual actions. As a result of that meeting, a plan was implemented to derive the best approach for providing defensible time margins.

The basic idea was to identify a time margin (or margins) for fire-related operator manual actions to ensure that they would be successful a very high percentage of the time (i.e., there is a high confidence of a low probability of failure). In other words, if analysts show in a demonstration that a randomly selected, established crew can successfully perform the actions, and show that the actions can be performed within a timeframe that allows for adequate time margin to cover potential variations in plant conditions and human performance, then the operator manual action would be shown to be reliable. For example, as long as analysts can show there is an "X-percent" time margin to perform a particular set of operator manual actions (e.g., the actions are shown during the demonstration to take less than 15 minutes, but even if they were assumed to take 30 minutes (or 100-percent time margin), plant damage or an undesirable plant condition will still be avoided) and all other important factors have been addressed, then they can be confident that the actions can be done reliably. Another approach may be to add a prescribed time (e.g., "Y" minutes) to the time obtained in a demonstration of any actions as a means to produce the desired increase in reliability.

---

[1] *Italicization* is used for emphasis.

The use of the time margin concept involves the derivation of appropriate time margins and a technical basis to support them. While the best technical basis would be empirical data from which the time margins could be derived, a database search was unable to find relevant data that could be used directly or generalized to the operator manual actions of interest. One potential exception was American National Standards Institute/American Nuclear Society (ANSI/ANS) Standard 58.8 [Ref. 2], which addresses time response design criteria for safety-related operator actions. However, it was determined that the data in ANSI/ANS 58.8 relevant to operator manual actions were limited and too broad to generalize well, they were probably overly conservative for most of the types of fire-related operator manual actions being considered, and they lacked clear and sufficient technical basis for our purposes.

Note that just one time margin was not necessarily being advocated; that is, the time margin could vary with the fire scenario, such that different margins may apply to different cases, regardless of whether the margins are measured in absolute (e.g., minutes) or relative (e.g., percent) time. Since varying time margins would most likely depend upon considerations such as fire frequency, magnitude, and consequences, this could be viewed as a form of "risk-informing" the criteria.

Thus, it was decided that an expert panel would be convened and that a facilitator-led, expert judgment process following the direct numerical estimation approach discussed in NUREG/CR-2743 [Ref. 3] and NUREG/CR-3688 [Ref. 4], in conjunction with the information and examples found in NUREG/CR-6372 [Ref. 5], would be used to identify reasonable time margins. The premise is that experts in the areas of nuclear power plant safety, risk assessment, inspection, fire safety and analysis, fire-related plant operations, human factors, and human reliability analysis could, in the context of a structured expert opinion elicitation process, make reasonable estimates of appropriate time margins.

B.2 First Expert Elicitation Meeting

A panel of six experts met at the NRC in Rockville, Maryland, on April 1 and 2, 2004. One week prior to the meeting, each expert was provided with a description of the goals of the meeting, which discussed many of the issues that would be addressed to generate the desired time margins.

B.2.1 Expert Panel and Qualifications

The six experts were as follows:

(1)   a Team Leader, Plant Engineering Branch, Division of Reactor Safety, in Region IV of the NRC; also serving as a project manager and inspector (covering plant engineering and maintenance) for the NRC over the past 14 years

(2)   a Reliability and Risk Engineer in the Probabilistic Risk Analysis Branch in the NRC Office of Nuclear Regulatory Research (RES); formerly a principal engineer (supervisor) and senior reactor operator at a commercial nuclear power plant licensee

(3)   a Senior-Level Advisor for probabilistic risk assessment, Division of Systems Safety and Analysis, NRC Office of Nuclear Reactor Regulation (NRR); formerly a project manager in the Energy Risk and Reliability Department at a contractor for the nuclear power industry

(4)   a principal of an independent contracting firm, especially contracting to Sandia National Laboratories, and recognized expert in the probabilistic analysis of fire and flood risk for nuclear and nonnuclear facilities; also a published author of numerous articles on this subject

(5)   an Engineering Psychologist in NRR/NRC with expertise in the area of human factors for more than 20 years; also serving as an NRC human factors expert on a national standards development committee in the area of human reliability analysis

(6)   a Senior Operations Engineer in NRR/NRC; formerly an NRC inspector for 20 years, starting as a region-based construction and fire protection inspector and including 8 years as a resident and senior resident at pressurized-water reactors (PWRs)

B.2.2 Summary of Topics Discussed During the First Meeting

Much of the first day, the discussion among the expert panel members and other meeting participants from NRR, RES, and RES contractors, including the elicitation facilitators, covered the following five topics:

(1)   What is this expert opinion elicitation all about?

(2)   What are the operator manual actions for which we are considering time margins?

(3)   What are the human performance influences that should be accounted for by the time margins?

(4)   What empirical data or other expert knowledge or experience may be relevant to developing the time margins and their bases?

(5)   How will the elicitation process work?

B.2.2.1 What Is this Expert Opinion Elicitation All About?

With regard to topic 1, it was agreed that the overall goal was to derive time margins that would provide reasonable assurance that local operator manual actions in response to fire, in general, can be achieved with a high confidence of a low probability of failure (e.g., 95-percent confidence of a 0.01 failure probability). While it was thought that specific numerical goals on confidence and probability were not practical, the experts were easily able to understand the intent of what we wanted to achieve. Further, so that all the experts' conception of the time margin was the same, the "model" shown in Figure B-1 was agreed upon as generally representative of the time margin concept.

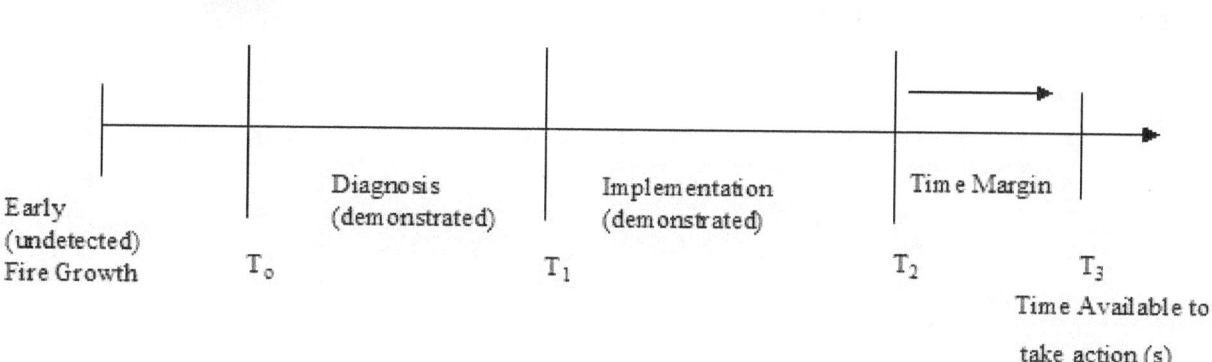

Figure B-1. Conceptual illustration of a time margin

B.2.2.2 What Are the Operator Manual Actions for Which We Are Considering Time Margins?

There was much discussion on topic 2. In particular, while it was agreed that we were addressing local (ex-control room) operator manual actions, there was confusion as to whether only preventive actions were included or whether reactive actions were also included. Further, there were clearly some differences in opinion as to when an action is a "repair." Preventive actions are those which, upon entering a fire plan/procedure, the licensee expects (without needing further diagnosis) to take to prevent spurious actuations or other fire-related failures so that adequate equipment is protected and hot shutdown can be achieved. Reactive actions constitute those taken during a fire in response to an undesired change in plant status and for which there is more of an element of detection of the undesired plant status and a diagnosis, with the aid of procedures, as to the correct actions to be taken. Further, there is precedence that repairs not be allowed for achieving hot shutdown.

While the expressed differences were not completely resolved, it was agreed that, in general, the following types of actions were *illustrative* of the types of actions we were concerned about:

- pulling fuses

- disconnecting power leads

- performing breaker manipulations (e.g., tripping, opening drawers, closing, changing switch positions) related to buses as well as individual loads such as valves, pumps, fans

- opening/closing/throttling of valves (e.g., with local switches, governor devices, handwheels)

- starting/stopping equipment, such as pumps and fans by either local switches/pushbuttons or breaker control

- installing jumpers or temporary power cables

- verifying or monitoring plant equipment or parameter status (and taking other actions as may be necessary based on these monitoring activities)

It was not the intent of this panel to define specifically what actions would or would not be allowed per the rulemaking that was in progress at the time. Therefore, the list above should not be construed as a list of what would have at that time been deemed "acceptable" operator manual actions. Nonetheless, it was agreed that the list was useful to generally define the typical kinds of actions for which time margins were to be considered, and that at least for purposes of the elicitation, both preventive and reactive actions would be addressed.

B.2.2.3 What Are the Human Performance Influences That Should Be Accounted for by the Time Margins?

With regard to topic 3, a number of observations were made. First, the rulemaking staff offered the following suggestions for the criteria:

- It should perhaps be made clear that the Available Indications criterion includes those indications necessary to detect and diagnose the location of the fire.

- It should perhaps be made clear that the Staffing and Training criterion allows both operators and maintenance staff to be involved as long as they are trained to take the desired actions.

- It should perhaps be made clear that the Communications criterion not only specifies that the communications systems must be adequate, but also that they must be readily available.

- It should perhaps be made clear that the Portable Equipment criterion specifically notes that such equipment includes what would be commonly referred to as "tools," such as keys, ladders, flashlights, gloves, and that these should be "staged" so that their locations are known and constant.

- It should perhaps be made clear that the Procedures criterion requires the use of *controlled* procedures.

- It should perhaps be made clear that, when multiple procedures will be required to be used simultaneously during a real fire (e.g., emergency operating procedures (EOPs) and the fire procedures), their simultaneous use will need to be part of the demonstration of operator manual actions in response to fires.

The staff offered these suggestions because it was clear that, in order to reasonably bound what the time margin was to account for, it was desirable that the other criteria be as specific and encompassing as possible. In this way, the time margin did not have to address potential inadequacies in meeting the other criteria and could focus on just those likely differences between what is expected in a typical demonstration of the actions vs. what might be experienced in a real fire situation (this became the basic premise for the time margin).

With this basic premise for the time margin, the discussion further elaborated upon what the time margin needed to take into account. Three possibilities were considered:

(1) The time margin should account for what an analyst is not likely to be able to recreate in the demonstration that could cause further delay (i.e., where the demonstration falls short), including the following examples:

- Random problems (i.e., not related to the fire) with instruments, indications, or other equipment such as a stiff handwheel or faulty communications device should be included.

- Environmental and other effects not easily included in the demonstration, such as smoke and toxic gas effects, increased noise levels due to the fire (e.g., alarms), water on the floor, fire hoses in the way, or too many people getting in each others' way, should be considered.

- Limitations of the demonstration to account for (or envelop) all possible fire locations where the operator manual actions are needed, resulting in different travel paths and distances to these locations, should be evaluated. (A similar limitation concerns the location and activities of needed plant personnel at the time the fire starts that could delay their participation in executing the operator manual actions, e.g., they may be on the opposite side of the plant and may need to restore certain equipment before being able to participate).

      –   Inability to execute relevant actions during the demonstration because of normal plant status or safety considerations while at power should be included.

(2)   The time margin should account for the fact that fire and related plant conditions can vary (e.g., fast energetic fire failing equipment quickly vs. slow-developing fire with little or no equipment failures for some time, variable fire detector response times and sensitivities, variable air flows affecting the fire and its growth, specific fire initiation location relative to important targets, presence (or not) of temporary transient combustibles, possible communication problems in some fires or in some noisy areas).

(3)   The time margin should account for the typical variability in human performance among individuals and among different crews and for the effects of human-centered factors that could become relevant during fire scenarios, such as stress, issues related to human factors and ergonomics (e.g., height at which task is performed), time pressure, and fear of fire, including the following examples:

      –   physical size and strength differences

      –   cognitive differences (e.g., memory ability, cognitive style differences)

      –   emotional response to the fire/smoke

      –   response to wearing a self-contained breathing apparatus (SCBA) to accomplish a task (i.e., some people may be very uncomfortable with masks over their faces)

      –   individual sensitivity to real-time pressure

      –   team characteristics

Further, it was agreed that these items did need to be part of the time margin for the following reasons:

- They address likely shortcomings of the demonstration (e.g., operators may not actually do the demonstration while wearing SCBAs or they may not perform the demonstration with full replication of environmental conditions, such as propagation of water on the floor into the rooms where the actions are to take place as a result of suppression system actuation in the room with the fire). *It was felt such shortcomings could result in potentially significant differences between times for actions during a demonstration and the times during real fires.*

- The demonstration can attempt to replicate only a small subset of all possible fires and resulting variability in fire and plant conditions (see examples cited under item 2 above), some of which could be worse than assumed in the demonstrations. *It was felt such variability could result in potentially significant differences between times for actions during a demonstration and the times during real fires.*

- It was recognized that some degree of human performance variability is to be expected, some of which could further delay the times to perform the desired actions during real fire situations. *It was felt such variability needed to be estimated and included in any derivation of time margins.*

Beyond this, it was agreed that the illustrative influences provided below, considering the categories mentioned above, were indeed representative of the influences that should be accounted for in the time margin:

- wearing SCBAs to complete the actions, which could affect performance in many ways, including the ability to communicate, etc. (use of SCBAs is not explicitly addressed by the rule criteria)

- substantial amounts of water on the floor from fighting the fire

- visibility problems due to smoke that is worse than assumed for the location of a given set of actions

- individual differences in the psychological effects of having to perform actions in proximity to a fire (even if the fire is not, in reality, physically threatening)

- inability to perform all of the subactions related to an "action" during a demonstration (e.g., the plant was "at-power" during the demonstration and certain actions could not be completely conducted while maintaining safety)

- time pressure (not sensed during demonstrations)

- the presence of less experienced staff, even though trained

- the need to identify alternate routes to and from the location of the operator manual actions because of the fire and its effects

- unexplained or unexpected equipment problems (e.g., a stuck handwheel, failures in communication equipment, misplaced tools, loss of lighting, loss of instrumentation)

- shortcomings in training not revealed during the demonstration

- inaccuracies in procedures for certain unique situations not previously identified (i.e., simply not thought of and not detected during the demonstration because the actual process could not be fully conducted)

- cases in which the fire is larger than expected and less time is available

Further, it was agreed that there could potentially be delays in either or both the diagnosis and decision to execute operator manual actions in response to fire as well as in the implementation of the desired manual actions; hence both effects should be considered when deciding on appropriate time margins.

While there was some discussion about how the analyzed time available (T_3) could be ascertained when it cannot be precisely known when a spurious or other fire-induced failure might occur, those discussions are not reproduced here since it was agreed that concerns about the appropriateness of T_3 (particularly as related to how to measure the time available for preventive actions) were not critical to the specific task before the experts. That is, determining the relevant time margins does not depend on the calculation of T_3.[2]

---

[2]   But the time margin is certainly relevant when evaluating whether the operator manual actions satisfy the timeline determined by T_3.

B.2.2.4 What Empirical Data or Other Expert Knowledge or Experience May Be Relevant to Developing the Time Margins and Their Bases?

Regarding topic 4, literature searches of easily available sources (only a short timeframe was available prior to the first elicitation) were performed in preparation for this meeting to seek any additional information that may be helpful to establish defensible time margins. Unfortunately, little was found. The following observations are provided to the extent they may be useful, but none of them are directly relevant to how to derive an appropriate time margin.

Actual events, recent inspections, and analytical processes suggest that, in spite of attempts to anticipate actual fire conditions and their effects, and then provide procedures, training, tools, communication devices, etc., so as to be able to perform the necessary or desired actions within expected time periods, the times to actually take the actions are often longer than prejudged estimates. The panel was prepared to discuss as many examples of this as may be desirable during the meeting. In some cases the difference between the actual time to perform the actions and the estimated time to take the actions has been small.

However, in extreme cases, as high as a threefold increase has been observed (i.e., it was estimated the actions could be taken within 30 minutes and the somewhat realistic time from a demonstration took nearly 90 minutes) for complex actions such as aligning, starting, and controlling a whole train of an injection system. In NUREG/CR-1278 [Ref. 6], it is noted that judgmental estimates are often low compared with actual times and that *a factor of 2 difference should not be unexpected.*

The above observations should be moot from our standpoint since the actions and their execution times are assumed to be obtained using the demonstration information. That is, the differences between judgmental estimates and times from the demonstration should not be an issue. Nonetheless, the above findings indicate that there may be time-delaying factors that are difficult to foresee, especially when other things can (and often do) go wrong. Thus, to the extent that the times from the demonstrations are still not entirely representative of all relevant actual fire situations (and demonstrating the actual times may be difficult, if not impossible, to achieve), it should not be surprising that the real times may still be even longer than what is obtained in a demonstration.

It was also observed that with regard to assessing risk significance, NEI-00-01 [Ref. 7] cites potential types of scenarios that should not be screened out as unimportant during the preliminary screening step of the information. Such a scenario includes one involving operator actions where both time is short (less than 1 hour) and the estimated time to perform the actions is greater than 50 percent of the available time. While not directly useful to deriving a defensible time margin, this step does seem to recognize that there may be factors that could make the time to perform the actions longer than estimated. The guidance implies that *a factor of up to 2 increase is desirable between the estimated time and the available time in order to provide adequate comfort that the actions can easily be performed in the available time.*

For the same reasons as cited earlier, this observation was not directly helpful as to how to derive a defensible time margin for action times obtained from a demonstration; however, it did support the idea that there are probably factors that can delay action times. Thus, a time margin is desirable to ensure that the actions can be reliably implemented.

B.2.2.5 How Will the Elicitation Process Work?

With regard to topic 5, the following process was used as initial expert opinion elicitations were performed on some sample cases:

- The facilitators summarized the relevant characteristics for which the time margin was being elicited (particularly, the types of actions and any relevant contexts for which the time margin applies, the relevant influences to be captured by the time margin, other applicable knowledge, experience, data, etc., and the form of the time margin). This was done in a facilitator-led discussion allowing experts to clarify these characteristics as necessary.

- Each expert privately estimated an appropriate recommended time margin.

- The experts' time margins were shared among the group and the experts were given the opportunity to provide their rationale for their estimates in a facilitator-led discussion. This identified legitimate considerations that were not accounted for by some experts, and it uncovered considerations that should not have been included by other experts. In either case, the results of the discussion caused some experts to provide a revised estimate.

- The experts were given a second (final) opportunity to privately arrive at a revised time margin.

- While we strove to reach a consensus on the identified time margins, the final elicited time margins from the experts were recorded and, as feasible, subsequently treated in a statistical manner to arrive at a single recommended time margin. (Following the completion of both expert opinion elicitation sessions, the facilitators decided that a strict statistical analysis was not warranted based on the limited results.)

Notes were taken during the entire meeting to subsequently and properly document the meeting's key discussions and decisions.

To support the experts in determining how best to derive their estimates of appropriate time margins, to help them decide what the forms of the time margins should be, and to determine how many different time margins were needed, the experts agreed that it would be helpful to consider a few sample operator manual actions and associated scenarios. The general goal was to see what could be learned by thinking about specific examples. From trying to determine appropriate time margins for a couple of specific cases, the experts thought they might be able to see trends, improve their understanding of the issues, and draw some general conclusions about time margins. In addition, it was proposed that, by examining specific cases of the types of fire operator manual actions being addressed and by considering the different types of influences thought to be important, the panel would better understand the nature of operator manual actions in response to fire and the ways in which the different influences might affect crew performance.

With these thoughts in mind, and with the remaining time available for the meeting, expert opinion elicitations were conducted on two example cases.

B.2.3 Example Elicitation Cases Addressed at the First Meeting

Two scenarios and related actions and timing were described to the experts for the example elicitation. One involved a preventive action that would be initiated as soon as the fire was detected, while the other was a reactive action that would be diagnosed on the basis of plant symptoms and relevant procedures. However, the cases were similar in that they both concerned the inappropriate opening of power-operated relief valves (PORVs) as a result of the fire. This is an important issue because the unexpected opening of the PORVs in a PWR can result in a significant loss-of-coolant accident (LOCA).

B.2.3.1 First Scenario/Action Case

In the first example scenario, a fire starts in an area that has the potential to cause inappropriate opening of the PORVs. Per the procedure associated with a fire in this area, once the fire is detected and located, a plant equipment operator (PEO) is summoned to the main control room (MCR) if necessary (although PEOs generally report to the MCR when events such as fires occur), provided with the relevant procedure, and directed to travel to the correct cabinet, find the correct terminal block, and pull the appropriate fuses to prevent the PORVs from opening. The PEO was assumed to then need to inform the MCR crew to provide verification that the PORVs were deenergized.

For purposes of the exercise, it was assumed that, during the plant's demonstration of this fire-related operator manual action (actually a set of subactions), likely fires in this area would normally be detected and located within approximately 5 minutes. Since by procedure the presence of the fire indicates the need for the appropriate fuses to be pulled, it was assumed that under most conditions the diagnosis for the need for the actions and the retrieval of the relevant procedures would be made in the same timeframe. Thus, T_1 was assumed to take about 5 minutes.

With respect to the time to execute the operator manual actions (T_2), it was assumed that the demonstration conducted at the plant revealed that a randomly selected, established crew accomplished the actions within about 4 minutes. That is, the responsible MCR person assigns a PEO and gives him the relevant procedure and instructions (about 1 minute), the PEO travels to the appropriate cabinet (1 minute), identifies and pulls the relevant fuses (1 minute), and notifies the MCR that the action was completed (1 minute), for a total of 4 minutes. (The experts at the meeting (including a former operator) agreed that this was a reasonable estimate of the time necessary to complete such an action for many plants.) The analyzed time available to complete the action before a problem would occur (T_3) was assumed to be approximately 20 minutes.

Given this scenario, it was the experts' job to identify and consider the factors that might delay performance of this task under realistic plant fire conditions. Per the guidelines discussed above, it was assumed that all of the operator manual action criteria had been met by the plant.

For this initial exercise, the panel members considered the three influence factors from Section B.2.2.3 of this appendix, focusing mainly on the factors that might not be covered adequately during the demonstration (i.e., aspects of the rule criteria that would not be easily addressed during the demonstration and could cause delays if problems arose). However, and especially during their modified responses, the experts also considered variations in plant conditions and human-centered factors in determining their time margins.

Table B-1 displays the increases in the time that were suggested by the experts to account for factors that might not be covered completely by the demonstration, as well as potential variability in plant conditions and fire scenarios and additional human influences. The suggested time increases cover factors that could reasonably delay the performance of the preventive actions associated with pulling fuses to prevent the PORVs from inadvertently opening due to the fire.

Table B-1 Initial and Revised Additional Times Added to Combined T_1 and T_2

Panel Member	Increase (Added to Original 9 min)		Factor (Total Time to Original 9 min)	
	Initial Estimate	Revised Estimate	Initial Estimate	Revised Estimate
#1	23 min	10 min	3.5	2.1
#2	6 min	10 min	1.7	2.1
#3	11 min	12 min	2.2	2.3
#4	6.5 min	9 min	1.7	2
#5	30 min	18 min	4.3	3
#6	1 min	10 min	1.1	2.1

A review of Table B-1 reveals a significant amount of variability in initial estimates of the amount of time that should be added to T_1 and T_2 to account for uncovered influences. After the panel members had the opportunity to discuss their results and share their reasoning with one another, much closer agreement was reached and, for the most part, the expert panel was converging on a factor of approximately 2 as an appropriate time margin for this case. That is, if it were assumed that the time to pull the fuses to prevent the opening of the PORVs might be twice as long as was obtained in the demonstration and still fall within T_3, then it would be appropriate to credit the action. In this case, since T_3 was assumed to be 20 minutes, and increasing the original time from the demonstration of 9 minutes by a factor of 2 results in a total of 18 minutes, then the reliability of the action would be shown.

However, it should be remembered that, as discussed at the end of Section B.2.2.5 of this appendix, the goal of the exercise was to see what could be learned by thinking about specific example cases. It was hoped that the exercise would support the experts' determination of how best to derive their estimates of appropriate time margins, to help them decide what the forms of the time margins should be, to familiarize them with the different types of influences thought to be important and how to consider their effects, and to determine how many different time margins might be needed.

B.2.3.2 Second Scenario/Action Case

The second scenario and action case examined at the meeting essentially served the same purpose as the first. That is, the goal was to continue to familiarize the panel members with the process and the factors to be considered to identify reasonable time margins for operator manual actions in response to fire.

For the second example (as with the first), the scenario involved a fire that starts in an area with the potential to lead to inappropriate opening of the PORVs. However, in this case, it was assumed that there is a reliance on a reactive process to deal with the potential opening of the PORVs. That is, the crew waits until there are some indications that the PORVs have opened, and then they send personnel out to pull the fuses to allow the PORVs to close (as a backup to the likely attempted closure of the PORV block valves).

For purposes of the exercise, it was once again assumed that it would take approximately 5 minutes to detect and locate the fire. In addition, it was assumed that another 2 minutes would pass before the fire caused the PORVs to open. Once the PORVs opened, it was assumed that the plant was able to show in the demonstration that diagnosis of the presence of the opened PORVs and contacting personnel to perform the needed actions could be done in about 1.5 minutes. Moreover, as in the preventive case, 3 minutes were assumed to travel to the cabinet, pull the fuses, and verify completion of the task with the MCR. Thus, in this case it was assumed that 4.5 minutes would be necessary to diagnose the need for the actions and to complete them, such that $T_1 + T_2 = 4.5$ minutes for the reactive case.

A difference between the reactive case and the preventive case is that the detection and location of the fire is not part of the assessment of the time margin.[3] Since the time between the start of the fire and the opening of the PORVs can be quite variable, the plant will be concerned with ensuring that, regardless of when the PORVs open, the PORVs will be closed in time to prevent any serious damage. Thus, the analyzed time available (T_3) is the worst-case time between the opening of the PORVs and the point at which serious damage would occur.

The only time that the activities associated with detecting and locating the fire would be relevant in the reactive case would be when the PORVs opened within the first 5 minutes after the fire starts. However, for this example it was assumed that the PORVs did not open until 2 minutes after the fire was located and detected. Thus, the panel focused on how much time they would need to add to the 4.5 minutes of T_1 and T_2 in order to account for the three influence factors discussed in Section B.2.2.3 above.

However, two caveats are relevant to this second example exercise. First, only a short period of time was available at the end of the second day of the elicitation session to perform the exercise, compelling the expert panel members to rush their judgments somewhat. Furthermore, based on discussions with the panel members, at least some did not agree that, for the case we were addressing, the activities occurring before the PORVs opened would not be relevant to the crew's performance in diagnosing the open PORVs and ensuring their closure by pulling the fuses. Thus, some panel members included adjustments to the fire location and detection phase and added that to their time adjustments, while others did not. Due to the limited time available for this example exercise, it was not possible in all cases to separate these extra time additions from the panel's estimates. In addition, there was not time for the panel to revise their initial estimates.

Table B-2 displays the increases in the time that were suggested by the experts to account for factors that might not be covered completely by the demonstration, as well as potential variability in plant conditions and fire scenarios, and additional human influences. The suggested time increases cover factors that could reasonably delay the performance of the reactive actions associated with pulling fuses to allow the PORVs to go closed before serious damage occurs.

---

[3]   Note that not all the panelists dismissed this time as irrelevant and included time margins in their overall assessment to account for influences that could arise during this specific interval.

Table B-2 Initial Time Added for Diagnosing the Need and Successfully Closing Open PORVs

Panel Member	Increase (Added to Original 4.5 min)	Factor (Total Time to Original 4.5 min)
#1[4]	13 min	2.1
#2	7.5 min	2.7
#3	7.5 min	2.7
#4	7.5 min	2.7
#5	25 min	6.6
#6	8.5 min	2.9

Despite some potential confounds with this example as discussed earlier in this section, it is worth noting that several experts were fairly close in their estimates. Based on the discussions with the expert panel members and the results above, it was considered possible that the time margin for reactive operator manual actions could be higher than for preventive actions.

B.2.4 Conclusion from First Meeting

As a result of the meeting, considerable insight was gained into reasons why it may be necessary to add a time margin to demonstration times and how large that time margin may need to be. At the end of the meeting, it was agreed that an additional elicitation meeting was necessary to pursue other representative examples of scenarios and actions to further learn what time margins would be appropriate for local operator manual actions in response to fire.

B.3 Second Expert Elicitation Meeting

The same panel of six experts (described in Section B.2.1 above) participated in the second expert opinion elicitation session held at the NRC in Rockville, Maryland, on May 4 and 5, 2004. Approximately 2 weeks prior to the second meeting, each expert was provided with a summary of the first meeting and given the opportunity to review the report, verify its contents (in particular the results of the example expert opinion elicitations), and make recommendations for changes. All panel members concurred with the summarized results of the first meeting as presented. In addition, a few days prior to the second meeting, an agenda for the second meeting was sent to the expert panel. The agenda noted the general steps planned for the meeting, reviewed important results from the first meeting, discussed the goals of the second meeting, outlined outstanding issues related to the time margins still to be addressed, and provided initial discussions of two possible examples for the second meeting.

---

[4] Panelist 1 added time for fire detection and location as well as to diagnosis of the open PORVs. Thus, the 13 additional minutes were compared relative to a total original time of 11.5 minutes rather than 4.5 minutes.

B.3.1 Summary of Topics Discussed During the Second Meeting

In the first meeting, two general types of local operator manual actions in response to fire were addressed and issues associated with the two types were discussed. The two types were preventive and reactive actions. Because some panel members and the facilitators had given additional thought to these types of actions since the last meeting, it was decided that the second meeting would begin by returning to a discussion of these types of actions.

B.3.1.1 Preventive Actions

It was repeated that for the preventive actions, it is generally assumed that once the fire has been detected and located, per procedure, the MCR crew directs someone to execute a number of actions that will prevent fire-related damage to equipment to ensure its availability to achieve its function during the fire scenario. Also by procedure, the only criterion for initiating these actions is the presence of the fire itself. However, in reality it is possible that crews may delay initiation of the actions for some period just to make sure that the fire is significant enough to initiate the actions. Moreover, it may take time for the appropriate crew member to retrieve the relevant procedures and assign plant personnel to complete the actions, etc.

During the second meeting some additional points were discussed about the preventive actions relevant to crediting them under the proposed operator manual action rule. First, it was noted that there are no guarantees that all preventive actions can be completed before the relevant equipment might be affected by the fire. There are many different kinds of fires in terms of initial size, growth rate, etc., and they can start in different locations within a room. Thus, while in many cases it may be relatively unlikely that a fire would spuriously affect equipment before the equipment could be protected by the operator manual actions, it is probably impossible to say that given actions can always be completed prior to the relevant equipment being affected by the fire. This being the case, it was argued that to take credit for such actions, it would need to be assumed that operators may have to perform reactive actions to restore the equipment to its functional state.

While panel members noted that plant procedures for preventive actions generally include steps to verify that the actions were successful, and if not, to take actions to ensure the equipment is placed in the appropriate state, they also noted that when demonstrating the feasibility of the actions and measuring the time it takes to complete the actions, these potential additional steps should be included. In other words, all preventive actions have the potential to involve reactive actions to ensure the availability of the equipment and, therefore, those additional steps should be included in demonstrating the actions and measuring the time to complete the action. The panel pointed out that while the resulting time estimates to complete the actions may be conservative for the cases in which the preventive actions are successful, if such aspects are included in the plant demonstration, then they should not have to be accounted for in the time margin.

The latter point became a critical aspect of the second expert elicitation meeting. The panel members argued that to be able to develop a reasonable time margin for operator manual actions in response to fire, the demonstrations of the actions should cover as many potential influences on performance as possible. Furthermore, the most reasonably conservative cases for the various conditions that could influence the ability of crews to complete the actions should be incorporated into the demonstration. In this way, the more extreme and less frequent variations in performance may be accounted for in the identified time margins, thereby making their development simpler and easier to justify.

It was argued that the appropriate range of conditions to be included in the plant demonstrations should be described. The result would be that the applicability of the time margins identified from this exercise would be contingent on plant staff demonstrating the actions as specified, for instance, in this document. Aspects to be included in the demonstration are discussed in Section B.3.1.4 of this appendix.

A final aspect about preventive actions discussed by the panel concerned how to measure the time to complete the actions (T_3). If there are at least some fire events that could affect important equipment before the preventive actions could be completed, then the time available to complete the actions (before serious equipment damage could occur and affect hot shutdown) should be measured from the earliest point at which the relevant equipment could be affected. Thus, if it is at all reasonable, analysts should assume that the fire could start exactly in the area where the equipment of concern would be affected at the earliest possible time. This may result in less time being available for preventive actions than might normally be assumed, which should be considered when analysts develop their timelines for operator manual actions in response to fires.

B.3.1.2 Reactive Actions

For the reactive actions, operators do not initiate the actions until they have detected and diagnosed that the relevant equipment has been affected by the fire and that it may be needed for hot shutdown. That is, they do not initiate the actions until the procedure, given the relevant indications, calls for the reactive actions. However, the panel noted that the symptoms indicating that the equipment has been affected could occur very early in the scenario when the crew is still in the process of detecting and locating the fire, entering initial EOPs, and possibly entering abnormal procedures. Alternatively, the symptoms could occur later in the scenario after the crew has been responding to the situation for a while and fire-specific procedures have been initiated. It was argued that, since the effect on the equipment could occur very early (e.g., as a result of an explosive switchgear fire), potential delays due to initial competing activities should be considered in determining the time margins. However, the panel was unable to conclude that the activities occurring during early stages of a fire scenario would necessarily be any more demanding that those occurring somewhat later in a scenario. It would seem that the demands of a given scenario across time would be plant- and scenario-specific; thus, this would be a factor that should be addressed by each plant for reactive actions, and the most reasonably conservative case with respect to potentially competing tasks should be modeled in the plant demonstration. If this is done, then any developed time margins would not have to take such effects into account.

The panel acknowledged that crews may find themselves dealing with "dueling procedures" at any point in a fire scenario and that the effects of possibly being in multiple procedures should be modeled to the extent possible during the demonstration of operator manual actions in response to fire.

Regarding the time available to complete reactive actions, T_3 would be determined by how much time would be available to restore the critical equipment after the fire effects had occurred in the context of the accident scenario.[5] Analysts should assess the worst case for when the effects could occur and calculate the time available on that basis. In many instances, it would seem that fire damage occurring as early as possible in the scenario would be the most serious (due to more time to build up to the expected high heat levels), but there may be some scenarios where this would not be the case. Again, analysts should consider such aspects in developing their timelines for the actions.

B.3.1.3 Other Types of Actions

Two other general categories of actions were considered by the panel. They included simple vs. complex actions and short-term vs. long-term actions. With respect to the latter, it was argued that essentially all local operator manual actions in response to fire would be relevant only in the short-term case (i.e., within the first hour of the scenario). Thus, it was decided that this distinction would not be relevant for developing the time margin.

However, over the 1.5 days of the meeting, the simple vs. complex distinction was discussed on several occasions. The issue was whether separate time margins would be needed for simple actions, such pulling a fuse, vs. more complex actions, such as multiple-task actions that involve coordination and communication among plant personnel. After examining the potential ways in which complexity might vary, it was decided that the nature of the specific actions being carried out by plant personnel would not vary significantly. That is, the actions being conducted by individuals would be of the general types of actions on which plant personnel are trained and perform routinely as part of their jobs. Thus, the complexity would more likely come from the coordination and communication associated with some activities and the associated time aspects.

The panel eventually concluded that, since both simple and complex actions would have to meet the same criteria in the (planned but discontinued) rule, and because time differences between tasks could be accounted for by using a common multiplier (e.g., a factor of 2 as a "time margin" multiplier on the demonstration) across all tasks, separate time margins as a function of complexity would not be needed. In fact, the panel eventually concluded that, *as long as all the (planned) rule criteria were met, the operator manual action demonstrations were performed appropriately (as described in the planned regulatory guide), and the time available for the various tasks was calculated appropriately, then a single time margin could be adopted.* The single time margin would cover all the remaining influences unaccounted for by the demonstration and could be applied generally to all types of operator manual actions in response to fire, including preventive and reactive actions. The influences on performance to be covered by the time margin and those to be covered by the demonstration are discussed below.

---

[5]   However, time zero would still be measured at initial fire detection, such that a plant using reactive type procedures would not necessarily have as much time to take actions as one with preventive procedures, due to the time delay between fire detection and initiation of operator manual actions.

B.3.1.4 Influences on Performance

Based on the results of the first meeting, the three influence factors listed in Section B.2.2.3 of this appendix were again assumed to be relevant to identifying an appropriate time margin. That is, it was thought that there were three factors that could lead to variations in the performance of the operator manual actions that would not generally be accounted for by meeting the rule criteria. Thus, it would be necessary to account for such influences in the time margin.

After further consideration of these sets of influences during the second meeting, the panel agreed that many of the aspects of the influence factors could be covered by assuming "worst-case" scenarios in both the conditions associated with a plant's demonstration of actions and in their calculation of how much time would be available to complete actions before serious equipment damage would occur and affect hot shutdown. As discussed above, such conservatism would limit the number of influence aspects that would have to be covered by the time margin.

The panel ultimately agreed that influence factor 2 (variability in fire and related plant conditions) should be addressed in the analyst's calculation of the time available for actions (T_3). Analysts should assume the worst-case reasonable variations in fire characteristics and plant conditions that could affect the time available to complete actions in that calculation. In addition, the panel agreed that some aspects of influence factor 1 (where the demonstration falls short) could be adequately addressed by making certain assumptions or simulating certain conditions during the demonstration. The demonstration should address the following aspects (among others):

- If it is reasonably likely that operators will wear SCBAs to complete actions, then they should wear them during the demonstration. Furthermore, if communication is necessary between operators under conditions in which they would wear SCBAs, then the communication should be achieved while wearing the SCBAs.

- If normal plant noise levels could affect communication in some areas, the demonstrations should be conducted under those conditions.

- If smoke could significantly affect visibility, then actions should not be credited.

- If it is possible that needed operator manual actions will involve plant personnel (e.g., PEOs) being summoned from other locations in the plant to obtain instructions and relevant procedures and proceed to the area of the actions, then the worst-case reasonable time for them to travel to the various locations, which may include traveling to the MCR, should be included in the time to execute the actions. In other words, in conducting the demonstration, necessary personnel should be located as far away as reasonable at the start of the simulation. In addition, the potential for such personnel to have to complete what they were doing before responding should also be considered in the demonstration and, therefore, in the time to complete the actions.

- If the fire or other factors could affect where personnel have to travel (e.g., what routes they have to take) and where they have to enter various rooms, then the worst-case reasonable effects should be modeled in the demonstration.

- If multiple actions (or multiple sets of actions) will have to be performed and coordinated and potential interference could occur, then all should be simulated in the demonstration.

The main point is that analysts should carefully analyze the potential context for given operator manual actions in response to fire and strive to model the worst-case, yet credible scenarios in their demonstrations. That is, they should do a good job of setting up their demonstrations to avoid being overly optimistic. For example, they should not select their most recently trained crew and then allow them to prepare for the demonstration (i.e., no "preconditioning").

B.3.1.5 Impact of Human Errors

Another topic of discussion concerned the impact of potential human errors in performing operator manual actions and the associated recovery actions. It was pointed out that, while the main goal of developing a time margin for local operator manual actions in response to fire was to cover the range of influences that could delay performance of the various actions, it is also possible that personnel could make errors in performing the actions. Although the probabilities of such errors may be relatively low, when they do occur, operators should be able to identify that an error has occurred and recover from the failure. Since verification is required for the operator manual actions (the proposed rule required that there be reliable indications available that actions have been completed), then it is reasonable to expect that the existence of any incorrectly performed actions or omissions could be detected. However, since it is probably not realistic to assume that analysts will model such recoveries in their demonstrations, the panel agreed that there should be at least some time built into the time margin to cover recovery actions (even if the likelihood of such errors occurring and not being caught immediately would be relatively low).

B.3.2 Determination of Time Margin

In order to determine an appropriate time margin, as in the first meeting, the panel thought that the process of stepping through reasonable examples of local operator manual actions in response to fire for estimating time margins was a useful exercise. By examining the various actions in some detail and thinking about how much delay could occur due to specific influences, it was thought that a good sense for a reasonable time margin would be obtained.

For this exercise in the second meeting, a somewhat more complex example of a preventive action (set of subactions) was addressed. This scenario was the third addressed across the two expert opinion elicitation meetings.

B.3.2.1 Third Scenario/Action Case

In this scenario, a fire starts in an area that has the potential to lead to inappropriate alignment or otherwise failure of the component cooling water (CCW) system. Per the procedure associated with a fire in this area, once the fire is detected and located, and in order to prevent CCW failure (the fire can supposedly affect all the equipment in Division A [Div-A] CCW, which is supposed to keep running, and the fire can potentially affect the Division B [Div-B] CCW valves, but not the Div-B pump, which does not start unless the Div-A train malfunctions), two PEOs are summoned to the MCR if necessary (PEOs generally report to the MCR when events such as fires occur). They are provided with the relevant fire procedure and are directed to travel to two locations; PEO 1 goes to the East Switchgear Room (ESWGR) and PEO 2 travels to the Div-B CCW room (the division to be protected). These rooms should not be affected by smoke from the fire, but the Div-B CCW room could, in a real fire, have a little water on the floor from nearby sprinkler operation if drains become partially plugged and some overflow occurs (this cannot be part of the demonstration).

Upon reaching their respective locations, PEO 1 is to communicate via radio with the MCR supervisor. The MCR staff then manually starts the Div-B CCW train and, after ensuring it is operating properly, the MCR staff shuts down the Div-A CCW train and pulls-to-lock the Div-A CCW pump. To protect the continued operability of the Div-B CCW train, PEO 1 is to pull three of many specifically labeled breakers (two breakers in one electrical cabinet at one end of the ESWGR and one breaker in a different cabinet at the other end of the ESWGR) that remove power from three Div-B CCW valves so they will stay in the proper position. PEO 1 is then to confirm with the MCR supervisor (via radio) that this is done and that Div-B CCW is continuing to adequately handle heat removal from the various loads. The MCR then informs PEO 2 (who has been listening in on his radio from the Div-B CCW room) that the Div-B CCW train is operating and that the manual crosstie valve between the CCW trains needs to be closed. PEO 2 then closes the manual crosstie valve in the Div-B CCW room and contacts the MCR and PEO 1 to confirm closure of the valve.

In the meanwhile, PEO 1 moves to the West Switchgear Room (WSWGR) and pulls the Div-A CCW pump breaker to ensure the pump cannot spuriously operate. PEO 1 then informs the MCR supervisor that the alignment is complete. The MCR supervisor verifies the alignment of the system via indicator lights, flows, and temperature indications and then releases the PEOs so they can attend to other matters.

Steps of the actions and times from the demonstration (or assumed times) are as follows:

Step 1.   For purposes of the exercise, it was assumed that, during the plant's demonstration of this fire and the operator manual actions, it was simulated that likely fires in this area would normally be detected and located within approximately 5 minutes.

Step 2.   Three additional minutes are expended for the PEOs to have reached the MCR and obtained the procedure and directions for the CCW manipulations (so now 8 total minutes have passed).

Step 3.   PEO 1 and PEO 2 reach their locations (travel time) and call in on the radios to ensure communication with each other and the MCR—4 minutes (so total time is now 12 minutes).

Step 4.   MCR staff starts Div-B CCW train, shuts down Div-A CCW train, pulls-to-lock the CCW A pump, and tells PEO 1 it is OK to pull breakers—1 minute (so total time is now 13 minutes).

Step 5.   PEO 1 pulls the breakers in the ESWGR and communicates with the MCR who ensure continued operation, and the MCR then informs it is OK to close the manual CCW valve—3 minutes (so the total time is now 16 minutes).

Step 6.   PEO 2 closes the manual valve and informs the MCR and PEO 2 of its closure—4 minutes (so the total time is now 20 minutes)

Step 7.   PEO 1 travels to the WSWGR, opens pump breaker, and communicates to MCR that this act is complete—3 minutes (so the total time is now 23 minutes).

Step 8.   MCR verifies all is OK and communicates to PEOs that they are released—1 minute (so the total time is now 24 minutes).

Table B-3 summarizes the expert panel's judgments for this scenario. In particular, the table shows the various steps of the actions being addressed, the time (assumed) for the actions obtained during the demonstration, and each panel member's judgment regarding what the total time for each step would be after adding time to account for various influence factors. Note that, at this point during the meeting, firm conclusions had not yet been reached regarding which factors should be addressed during the demonstration in calculating available time, as opposed to what should be included in the time margin. In fact, much of that information came out of discussions held during and after the scenario exercise. Which of the three general influences from Section B.2.2.3 above that the panel considered potentially relevant for each step of the action is noted in the table.

Table B-3 Total Time for Each Step of the Action for the Third Scenario, by Panel Member (Base Time Plus Time Added for Influence Factors)

Step and (Base Time)	Relevant Influence Factors	Panel Members' Total Times for Each Step (min)					
		#1	#2	#3	#4	#5	#6
1—(5 min)	#3	5	5	5	5	5	5
2—(3 min)	All	4	5	4	4	3	3
3—(4 min)	All	6	4	6	6	7	5
4—(1 min)	#1, #3	1.5	1	2	2	2	1.5
5—(3 min)	All	5	5	5	6	5	4.5
6—(4 min)	All	7	5	8	14	7	5
7—(3 min)	All	5	3	3	7	3	3
8—(1 min)	All	1.5	2	1	2	3	1
Total (24 min)		35	30	34	46	33	28

Each panel member considered how he or she thought the different influence factors might lead to increases in the time to complete each step of the action. A review of the table indicates that the total increases range from a factor of 1.25 to about 2, with an average of about 1.5, or an increase of 50 percent in the time. After the panel members had discussed the reasons for their additions, many thought that a factor of 1.5 to 2 might be a reasonable time margin for operator manual actions. However, they also recalled that, in working through the earlier examples, some panel members had identified greater relative time increases and had been considering significantly larger time margins.

B.3.2.2 Fourth Scenario/Action Case

By the time the fourth scenario was addressed, several discussions had taken place and the panel had agreed that influence factor 2 associated with fire characteristics and plant conditions should be addressed by analysts in determining the time available to complete the actions (as discussed in Section B.2.2.3 above). Similarly, the panel had identified several important factors that might lead to significant variation in performance that should also be addressed by analysts in conducting the demonstrations and noted that this should be made clear. Thus, in the final exercise, there were two major goals. One was to assess actions assuming the plant had performed a proper demonstration. The second was to address a preventive action that included the situation in which the equipment was affected by the fire before the preventive measures were completed, requiring the operators to perform the relevant reactive actions. The idea was that by addressing a hybrid, they would have the opportunity to assess a range of potential influences under conditions different from those considered before.

The example used was similar to that used for the third scenario, except that in this case, in addition to PEO 1 having to pull the breakers for the Div-B CCW valves in the ESWGR and communicating with the MCR and PEO 2, PEO 1 will have to travel to the relevant room and verify and check on the valve positions of the Div-B CCW valves and readjust as necessary. In this case, it is assumed that the Div-B CCW system has been affected by the fire and the operators enter a more reactive mode. For the exercise, it was assumed that three alignment valves in Div-B CCW have spuriously closed. PEO 1 will need to reopen the valves and take the steps necessary to restore flow.

The steps considered in the elicitation were the same as before (see Section B.3.2.1 of this appendix) with the following exceptions:

Step 5.   Normally, PEO 1 pulls the breakers in the ESWGR and communicates with the MCR crew, who ensure continued operation, and the MCR then informs PEO 2 that it is OK to close the manual CCW valve—3 minutes (so the total time is normally 16 minutes). However, now PEO 1 discovers that three of the valves have spuriously closed and need to be repositioned. PEO 1 needs to reopen the valves, restore flow to the Div-B CCW system, and inform the MCR—12 minutes added (so now the total is 28 minutes).

Step 7.   Deleted (small effect; limited time remaining to panelists).

Step 8.   Deleted (small effect; limited time remaining to panelists). For this exercise the scenario was ended after Step 6, so the total time was 32 minutes (previous 24 total minutes plus additional 12 minutes from Step 5 minus 4 minutes from Steps 7 and 8).

For this final exercise, the expert elicitation was done in a manner slightly different from the other examples. This was partially attributable to the limited time remaining on the second day; it was viewed as an approximate but expedited way to combine both the initial and revised estimation steps. In this case, each member decided how much time he or she thought needed to be added to each step of the operator manual action based on the influences, and the panel discussed the basis for the selected times among themselves. Finally, each member settled on a value he or she thought was reasonable and the facilitators documented the range of values proposed by the panel. In cases in which several panel members were in agreement about the values, the mode (most repeated value) was also identified.

Table B-4 presents the results of the final elicitation, displaying the times added by panel members from considering influence factors that could not be covered in the demonstration (influence factor 1 in Section B.2.2.3 above) and the times added by considering human-centered influences (influence factor 3 in Section B.2.2.3). As noted above, aspects associated with fire characteristics and plant conditions (influence factor 2 in Section B.2.2.3) were assumed to be addressed by the plant and were not covered in the example.

Table B-4 Time Added to Each Step of the Manual Action for the Fourth Scenario (Hybrid Case of a Preventive and a Reactive Action)

Step and (Base Time)	Influence Factor 1 (Demonstration Shortfalls)	Influence Factor 3 (Human-Centered Factors)
1—Fire detected and verified (5 min)	No time added	No time added
2—PEOs to MCR (3 min)	1 min (panel agrees)—minor smoke, obstacles, etc.	0.5–1.5 min
3—PEOs to remote locations (4 min)	1–2 min—minor smoke, communications delays	0.5–2 min
4—MCR starts CCW B train and stops the A train (1 min)	0.2–1 min—MCR activities (fire distractions)	0–0.5 min
5—PEO 1 initially pulls breakers (3 min)	0–0.5 min	1–3 min (mode = 1.5 min)
5a—PEOs 1 and 2 determine that three valves on Div-B CCW have already spuriously closed. Re-open valves and restore system (12 min)	2–6 min	2–3 min (mode = 3 min)
6—PEO 2 closes crosstie (4 min)	2–4 min (assumed water on the floor, or other conditions.)	1–3 min (mode = 2 min)
Total (32 min)	Total of 6.2–14.5 min added	Total of 5–13 min added

When the total time added for the two influences categories are combined, the range of times to be added to cover their impact is 11.2–27.5 minutes. When these times are added to the base times (in the first column), the range is 43–60 minutes, which once again would represent an increase in the base time of roughly 50–100 percent.

B.4 Identification of Time Margin and Conclusion

Based on their reviews of the influence factors, the results of the example elicitations, and the need to allow some time for potential recovery actions, the panel members agreed that *a time margin factor of at least 2 would allow for a "high confidence of a low probability of failure" for local operator manual actions in response to fire*. The implication at the time with respect to the rulemaking activity was that, as long as operating plants meet the rule criteria for the actions (address the appropriate factors), they perform sound demonstrations of the actions at the plant (as described herein), perform reasonable calculations of the time available for the various actions (information for which is discussed herein), and can show that the time available is at least 100 percent greater than the time obtained in the demonstration, then local operator manual actions in response to fire could be considered reliable.

B.5 Characteristics of an Expert Elicitation Panel

As noted in the introduction to this appendix, analysts may find the discussion of the expert elicitation process useful to their efforts associated with estimating the potential impact of the factors creating the uncertainties to ensure that there is adequate extra time. While analysts may prefer a different approach for evaluating the uncertainties associated with manual actions (particularly if they are relatively simple and plenty of extra time is clearly available), if an expert panel elicitation is used, this section provides a brief summary of the characteristics and types of expertise that would be appropriate for such a panel.

An expert elicitation panel is typically composed of independent specialists, recognized in at least one of the areas/specialties addressed by the topic under evaluation. Generally, a multidisciplinary team approach should be used. The technical disciplines involved may vary depending on the particular topic (e.g., fire scenario) being analyzed (i.e., if there is a radiation hazard associated with the fire scenario, a radiation protection specialist or health pPhysicist may be needed as part of the team). In general, the expertise required could include human reliability analysis, human factors, fire protection, operations, instrumentation and control engineering, training, procedure development, probabilistic risk assessment, and other expertise as indicated by the fire scenarios and actions being examined. The team's objective is to arrive at an estimate of the time margin necessary to envelop the uncertainties associated with manual actions through consensus of the members.

An expert elicitation panel has both advantages and disadvantages. The principal advantage of the panel is the participants' knowledge and expertise in the subject area. Additionally, the panel approach can provide significant reductions in time and cost allocations compared to other evaluation techniques, and leverage the credibility of conclusions because of the panel members' expertise. The tool also has limitations, significant among them is the elimination of minority view points because of consensus-based conclusions and the potential for the view of a "dominant" member to be overly influential in the decisionmaking process. Additionally, there is evidence that operators can sometimes be optimistic about action implementation times and such bias needs to be controlled [Ref 6]. References 5 and 8 provide information and other references on controlling for various sources of bias.

B.6 References

1. Forester, J.A., and A.M. Kolaczkowski, "Summary of Expert Opinion Elicitation on Determining Acceptable Time Margins for Local Operator Manual Actions in Response to Fire: Results of Initial Meeting Held April 1 and 2, 2004, and Final Meeting Held May 4 and 5, 2004," Letter Report to the U.S. Nuclear Regulatory Commission, Sandia National Laboratories, June 2, 2004.

2. American Nuclear Society, "American National Standard Time Response Design Criteria for Safety-Related Operator Actions," ANSI/ANS Standard 58.8-1994, La Grange Park, Illinois.

3. Seaver, D.A., and W.G. Stillwell, "Procedures for Using Expert Judgment To Estimate Human Error Probabilities in Nuclear Power Plants," NUREG/CR-2743, U.S. Nuclear Regulatory Commission, Washington, DC, 1983.

4. Comer, M.K., D.A. Seaver, W.G. Stillwell, and C.D. Gaddy, "Generating Human Reliability Estimates Using Expert Judgment," NUREG/CR-3688, Volumes 1 and 2, U.S. Nuclear Regulatory Commission, Washington, DC, 1984.

5. Budnitz, R.J., G.M. Apostolakis, D.M. Boore, L.S. Cluff, K.J. Coppersmith, C.A. Cornell, and P.A. Morris, "Recommendations for Probabilistic Seismic Hazard Analysis: Guidance on Uncertainty and Use of Experts," NUREG/CR-6372, U.S. Nuclear Regulatory Commission, Washington, DC, 1997.

6. Swain, A.D., and H.E. Guttman, "Handbook of Human Reliability Analysis with Emphasis on Nuclear Power Plant Applications—Final Report," NUREG/CR-1278, U.S. Nuclear Regulatory Commission, Washington, DC, 1983.

7. Nuclear Energy Institute, "Guidance for Post-Fire Safe Shutdown Analysis," NEI 00-01, Revision 0, Washington, DC, May 2003.

8. Forester, J.A., A.M. Kolaczkowski, S.E. Cooper, D.C. Bley, and E. Lois, "ATHEANA User's Guide," NUREG-1880, U.S. Nuclear Regulatory Commission, Washington, DC, July 2007.